The Philosophy of Global Warming

Dr Neil Paul Cummins

Copyright © 2014 by Neil Paul Cummins

All rights reserved. This book, or parts thereof, may not be reproduced in any form without permission.

A catalogue record for this book is available from the British Library

ISBN: 978-1-907962-99-8

Published by Cranmore Publications

Brighton, England

http://www.cranmorepublications.co.uk

This is my most comprehensive and definitive work. In it you will learn:

What the philosophy of global warming is and why it is of great importance.

Why the decision-making process concerning the appropriate human response to global warming requires a consideration of the evolutionary forces which propel the planet.

Why cutting fossil fuel emissions is a futile exercise.

What the human species is and how it relates to the non-human life-forms of the Earth.

Why the human species has a special place in the universe and how this is related to global warming.

What it means to say that your life has a purpose.

Why the evolution of technology and the evolution of spirituality are deeply interconnected.

Why there is an urgent need for the technological regulation of the temperature of the Earth's atmosphere.

Contents

The Purpose of This Book   9

Introduction   13

PART 1: PHILOSOPHY

1	What is the Philosophy of Global Warming?	23
2	The Two Paths Facing Humanity	28
3	Two Types of Global Warming	33
4	The History of Our Solar System	49
5	What is Life?	55
6	What is the Human Species?	62
7	Technology and the Environmental Crisis	73
8	Why Life Benefits From Technology	89
9	Is the Damage Already Done?	99
10	The Evolutionary Processes Which Propel the Planet	110
11	Humans in the Cosmos	130
12	The Interplay between Technology and Spirituality	145

PART 2: DIALOGUE

A plethora of objections, questions and queries relating
to my philosophical worldview are posed and answered   157

PART 3: ARTICLES

Was the Cosmic Bringing Forth of Humans 'Inevitable'?	225
Two Routes to the Need for Geoengineering	229
The Need for Geoengineering	231
The Nature of the Universe	234
Links between My Philosophy & the Buddhist Theory of Atoms	236
The GreenSpirit Journal Comments on ITHSS	240
The First Book Critiquing ITHSS	242
Ahead of the Curve	247
The Need for a New View of Humans in the Cosmos	250
Technology	254
Human Population & the Environmental Crisis	258

The Growing Realisation of the Need for Geoengineering the GMST	262
Humans and Other Animals	264
Animals Think like Humans	267
Earth 'Four Years from Disaster'	269
The Futility of Emissions Cuts	272
Prepare for Extreme Global Warming	275
Emissions Cuts: The Gap between Ambition & Reality	278
Accelerating Polar Ice Melting & Geoengineering	282
Evolution versus Creationism	288
The Calm before the Carbon Storm	292
Perceptions of Global Warming	298
Global Warming: Perceptions, Responses & Energy Policy	303
Global Warming & the Anthropocentric and Ecocentric Attitudes	309
George Monbiot on Atmospheric Carbon Dioxide Concentrations Reaching 400ppm	313
The Three Questions & the Philosophical Worldview	317
The Environmental Crisis & the Colonization of Space	325
Technology and Stewardship	328

The Inevitability of Geoengineering	332
The Conceptual Framing of Geoengineering	336
The Technological Healers of the Earth	338
The Concept of 'Future Generations'	346
Is Fracking Good or Bad?	351
Extreme Weather Events & Global Warming	355
How Much of Man is Natural?	360
Friedrich Hölderlin and the Environmental Crisis	373
Friedrich Hölderlin: A Final Reflection	403
Further Reading	406
Keeping in Contact	411

The Purpose of This Book

The purpose of this book is to get you to think about the philosophy of global warming. I am very hopeful that the information that is presented will change how you perceive the human presence on the Earth. I am hoping that you will conclude that the human presence on the planet is a positive one, a sign that the Earth, life, and even the Solar System, is positively thriving. I have three main reasons for hoping to convince you of this.

Firstly, I sincerely believe it to be true, and as a deeply philosophical person I simply have the desire to express the truth and to help other people to see the truth. You might be curious as to the source of my beliefs. Furthermore, you might be thinking, are my beliefs just my beliefs or are they 'the truth'? All I can really say on this is that the beliefs and views that I outline in this book seem to me to arise from an episode of direct personal insight which was backed up by subsequent knowledge acquired from the insight and work of others. I am not an expert on the phenomenon of direct personal insight, of personal revelation into the truths of the universe, but I believe that it is possible that the universe can directly endow individuals who are in a certain state (a state of 'receptivity') with certain truths about itself. Perhaps such an endowment was the catalyst for my move into academia in my mid-twenties. My childhood years were spent in the deepest depths of the Cornish countryside, surrounded by thousands of trees and very few people. In my mid-twenties I had been living on a very small island, which is situated in the Atlantic Ocean, for a number of years. Again, as with my childhood years, I was surrounded mainly by non-human nature, the powerful ocean waves, the sometimes fierce weather, the plentiful beaches and the wilderness. After several years of doing a menial, unfulfilling and soul-destroying job on this island something changed within me, some kind of awakening occurred. There arose within me a new sense of openness; I spent time just looking at my surroundings, really looking; things appeared slightly differently than they did before, more alive, more vibrant. Questions and insights bubbled up within me and I had little choice but to seek to follow their lead. These initial experiences and questions led to a journey of well over a decade; a journey that involved attaining a first class BSc in Environmental Studies, an MA in Philosophy, a PhD in Philosophy, an international writing prize, conference speeches in Venice and Marburg, and finally, this book.

Secondly, I am slightly concerned by the increasing dominance of the view that the human presence on the planet is a destructive one. This view increasingly pervades the media, the arts, culture, various academic disciplines, politics and even religion. I recently attended a conference where there were speakers from a variety of religions and I was surprised by what they said. Not a single speaker had anything positive to say about human existence; there was talk of environmental destruction, overpopulation, and it was even suggested that the theological talk of a special place for the human species on the planet (the view of human dominion) was a view that needed to be rejected. According to this increasingly dominant view humans are, at best, just one species among many, and at worst they are the despicable destroyers of life. This view concerns me because it has led to movements such as the Voluntary Human Extinction Movement (VHEMT) which was founded in 1991. If it is widely accepted that the human presence on the planet is a negative one, and that there are too many humans on the planet, then it seems increasingly likely that plans will be instigated to cull the human species; in other words, billions of people could ultimately be needlessly killed (I don't know exactly how this might be done, or who might do it, but I know there are people who think this would be desirable and who think about how it could be done; there are even people who think that it is already being done). Needless mass murder based on a false philosophy is something that I would like to see averted.

Thirdly, if the place of the human species on the Earth that I outline in this book is widely accepted, then a range of positive outcomes can result. We can celebrate our uniqueness, celebrate the joy that we are bringing to the Earth, rather than wallowing in despair at the thought that we are seemingly destroying the planet without really wanting to. Because states such as joy and despair ripple out from all sources where they exist, a more joyous philosophy would result in a more peaceful and joyous planet. We can also increasingly appreciate the value and perspectives of all individuals, all cultures, all perspectives, all life-forms, all personalities, as each of these has a positive role to play in the glorious evolutionary unfolding of the Earth. Furthermore, the realisation of our place on the planet, our purpose as a species, can enable us to reallocate our limited resources so that this purpose is more speedily fulfilled. Currently an enormous amount of resources are wasted on global warming mitigation schemes; these resources could be more optimally allocated. The creative energies of individuals can simultaneously be optimised. The outcome of this optimisation, through speeding up the fulfilment of our purpose, would be to more speedily bring about a more sustainable and harmonious existence, an increasingly peaceful and spiritual human presence on the Earth.

I have used a variety of writing styles, perspectives and approaches to present the information in this book. There are three parts to the book. Part 1 contains twelve chapters each of which contains a particular theme which is of relevance to the philosophy of global warming. Taken as a whole this part of the book can be thought of as providing a detailed overview of my philosophical worldview. Part 2 is a dialogue in which an objector to my philosophy poses a multitude of questions/queries/concerns and I provide responses. Part 3 contains a plethora of articles each of which illuminates certain aspects of my philosophy. The reason for this three-pronged approach is that what I am trying to get you to see is complex and it involves interconnections between many different phenomena. You are also likely to come across things which violently clash with your existing beliefs. My hope is that the three-pronged approach will both help you to understand particular points, and also to comprehend the bigger picture. You might find a particular chapter irrelevant at the time of reading it, but if you are open to the possibility that every chapter, every paragraph, is but a small jigsaw piece, then by the end of Part 3 you should be able to see the complete interconnected cosmic puzzle. There might be a complete transformation in the way that you see the world around you. In order to get the most out of the book I would definitely recommend starting at the beginning and moving through page by page, rather than jumping ahead to various sections that seem particularly interesting. I have attempted to slowly build up an overall philosophical worldview as the book progresses; that which appears in the latter stages of the book assumes an understanding of that which comes before.

Introduction

Do you believe in global warming? Do you believe that humans are the cause of this phenomenon? Do you believe that global warming poses a real threat to both humanity and to non-human life on the planet? I think it is safe to say that the majority of people would answer these questions as follows:

Do you believe in global warming?

Yes. Global warming is occurring because carbon dioxide concentrations are increasing in the atmosphere; this exacerbates the 'greenhouse effect' and causes global warming.

Do you believe that humans are the cause of this phenomenon?

Yes. Humans are the cause of global warming because atmospheric carbon dioxide concentrations have shot upwards since the start of the Industrial Revolution, as revealed by the 'hockey stick' graph. This has occurred because of the human use of enormous amounts of fossil fuels and also because of the human destruction of rainforests and other carbon sinks.

Do you believe that global warming poses a real threat to both humanity and to non-human life on the planet?

Yes. The polar ice will melt, sea levels will rise, the climate will significantly change, extreme weather events will become more pervasive, the food supply will be badly affected, temperature increases will make large parts of the planet (or even the entire planet) inhospitable; in short, the conditions which currently enable humans and non-human life-forms to flourish might disappear.

The reason that I think it is safe to say that the majority of people would answer these questions in such a manner is that these views are so pervasive in mainstream media, politics, culture and academia. These views, in turn, arise from the science of global warming. The scientific understanding of global warming centres on the 'greenhouse effect'. The 'greenhouse effect' is a natural phenomenon the existence of which is necessary for human existence; without it the atmosphere would be far too cold for humans to exist. The 'greenhouse effect' exists because greenhouse gases, such as carbon dioxide and methane, trap the incoming infrared radiation from the Sun after it has bounced off the surface of the Earth; this trapping warms up the Earth's atmosphere. The term the 'greenhouse effect' is often used to refer simply to the fact that by increasing greenhouse gas concentrations in the atmosphere humans have exacerbated this natural pre-existing effect, thereby causing a higher atmospheric temperature than would otherwise have been the case. It is useful to keep in mind that the 'greenhouse effect' is a non-human effect which has been affected by humans.

Introduction

The science of global warming has numerous dimensions. Scientific measurements have revealed the levels of greenhouse gas concentrations in the distant past through ice cores and tree rings, and they have revealed recent and current concentrations through direct measurement. Such measurements have produced the 'hockey stick' graph which shows escalating atmospheric carbon dioxide concentrations in very recent post-industrialisation times. In this period humans have removed 'carbon sinks' by engaging in mass deforestation, whilst simultaneously releasing enormous amounts of fossil fuels from their underground storage areas. Given these activities one should not be surprised that the measurements made by scientists have produced the 'hockey stick' graph. Scientists are also measuring the polar ice, measuring sea levels, and producing a plethora of computer models which attempt to predict how a warmer atmosphere will change the climate in various regions of the Earth.

The science of global warming is well established. I do not doubt the science of global warming. There are those who do doubt the science of global warming. Some people claim that the 'hockey stick' effect of recent escalating atmospheric carbon dioxide concentrations is caused by 'natural variation' rather than by human activities. There are others who accept that humans have caused the 'hockey stick' effect, but deny that rising atmospheric carbon dioxide concentrations lead to global warming. There are even a few people who deny that the 'hockey stick' effect reflects reality, believing that it has been created by the manipulation of data by scientists. There are almost always people with minority views. I myself am convinced by the science of global warming and thus believe that human activity has resulted in an increase in greenhouse gas concentrations in the atmosphere, and that such an increase leads to a warmer atmosphere through exacerbating the 'greenhouse effect'.

This book is not about the science of global warming; it is about the bigger picture, the wider situation within which the science of global warming is situated. This wider approach is needed because science has come to dominate the debate concerning global warming, and there are other non-scientific factors which need to be considered, factors which are of crucial significance. The initial domination by science of the global warming debate was inevitable; after all, we only know about the phenomenon because of scientific enquiry. However, the time has come to widen the debate, to widen our understanding of the factors relating to the phenomenon of global warming. The time has come to fully engage with the philosophy of global warming.

Of course, non-scientific factors have already been widely discussed concerning the phenomenon of global warming. The science of global warming has established that *there is a problem that needs to be addressed.* Since this scientific realisation occurred the phenomenon has inevitably encroached into the domains of politics, ethics, economics, psychology, business and engineering.

At the international level political leaders frequently meet to draw up protocols and to discuss how to respond to the problem. At the domestic level politicians seek favour with sections of the electorate by saying that they will respond to the problem. Environmental charities have taken up the cause and have sought their own solutions to the problem.

Introduction

In the realm of ethics, discussions take place concerning who is to blame for the problem and who should bear the consequences and financial cost of dealing with the problem; the rich countries might be the historical cause of the problem, but should poorer countries be prohibited from industrialising in the same fossil-fuel intensive way? Should rich countries provide less fossil-fuel intensive technologies to the poorer industrialising countries for the sake of everyone across the planet? What is the fair thing to do?

In the realm of economics there are discussions concerning how to get countries and individuals to have lower carbon footprints; we are here in the realm of taxes, subsidies, incentives and tradable permits. Psychologists hone in on the individuals and seek to understand how they can be made to use less resources, how they can change their lifestyles, how they can come to see the connections between their individual actions and the larger planetary problem of global warming. Businesses respond to the problem through presenting an 'environmentally friendly' carbon-neutral face in order to attract more custom; they also seek to come up with genuine solutions to the problem such as technologies to help humans cope with a changing and more hostile climate. And engineers are working on a plethora of solutions to deal with the problem; these range from enhanced sea wall defences to technologies to pull carbon dioxide directly out of the atmosphere so that it can be placed (back) in underground storage.

The science of global warming has clearly encroached into a wide range of disciplines. What has yet to occur is for the nature of the 'problem' itself to be seriously enquired into. The 'problem' itself is simply a scientific fact. It is a scientific fact that the temperature of the Earth's atmosphere is regulated by the 'greenhouse effect' and that human activity has resulted in an increase in greenhouse gas concentrations in the atmosphere (as I have already stated, I am convinced that this is a fact). However, a scientific fact such as this, a fact which presents a problem, doesn't automatically simultaneously present its own solution.

A simple way of looking at the situation would be as follows:

- **Scientific Fact** = Human activities have increased the carbon dioxide concentrations in the atmosphere.

- **Problem** = If the increase is of a sufficient magnitude global warming will occur to the detriment of both human and non-human life-forms.

- **Solution** = Humans need to stop carbon dioxide concentrations in the atmosphere from rising too much.

This is not only a simple way of looking at things, it is also surely true. However, the important point is that what exactly the solution to the problem entails is not clear. In other words, there are two ways in which humans *might* be able to stop carbon dioxide concentrations in the atmosphere from rising too much:

Introduction

>**Path 1**: Humans stop emitting, or radically reduce emissions of, carbon into the atmosphere.

>**Path 2**: Humans use technology to regulate the amount of carbon dioxide in the atmosphere.

I used the phrase 'might be able to stop' because according to one line of thought, a line of thought which is barely mentioned in the media, Path 1 is not even a possible solution. There are two different reasons why this might be so. Firstly, the damage has already been done, simply stopping now will have no effect; the action-consequence time-lags mean that carbon dioxide concentrations are set to keep on rising for the foreseeable future whatever we do now. Secondly, we simply cannot stop emitting now; the state of the world (population size and growth, economic trajectories, developing countries industrialising, state of technology) and the human dependency on cheap fossil fuel energy supplies means that Path 1 is nothing more than a pipedream, mere fanciful wishful thinking.

Despite this line of thought there is currently a widespread view which pervades the minds of most people – the politicians, the media, the activists, and the general public – that Path 1 is the solution to the problem. Despite the reality of the situation, which is carbon emissions continually rising across the world, and immense future changes already 'locked in' through action-consequence time-lags – Path 1 utterly dominates debates concerning the phenomenon. This seems to be the instinctive, almost childlike, response: if the problem is releasing carbon into the atmosphere, the solution has to be to stop releasing carbon into the atmosphere (Path 1).

The situation that we face is actually much more complex than is belied by this simple instinctive response. In other words, the question of which path humanity needs to adopt is a very complex question. The appropriate answer to the question requires a consideration of a wide range of both scientific and non-scientific factors. So, there is a scientifically-revealed problem which presents two possible solutions (Path 1 and Path 2). The discovery of the appropriate solution to this problem requires a careful consideration of a number of diverse factors, factors which have not yet been widely considered in relation to the problem. To find the appropriate solution we need to shift our focus from the science and delve deeply into the philosophy of global warming.

PART 1: PHILOSOPHY

Chapter 1

What is the Philosophy of Global Warming?

The science of global warming is wholly concerned with measurements and with numbers. In other words, it is concerned with measuring instruments, the numbers recorded by these instruments, and with data of other kinds. There are measurements for current atmospheric greenhouse gas concentrations, for past atmospheric greenhouse gas concentrations, for changes in polar ice cover, for sea level rise, for atmosphere-ocean interactions; there are also numerical projections for future emissions, for future greenhouse gas concentrations, and for the future temperature and climate in various parts of the planet.

The question of extreme importance is: Can measurements and numbers be a sufficient basis for a course of action? In other words, can one's personal actions, or the actions of the human species, appropriately be wholly grounded in measurements/numbers? It seems obvious to me that the answer is *no*. Numbers are useful but they cannot themselves determine an appropriate course of action. For example, a scientific measurement-derived number might be that there is a 95% chance of precipitation in the area in which I live. This is a useful number to know about, but it doesn't wholly determine whether I will take a particular course of action. In order to come to a decision about what course of action I will take a whole host of other non-scientific, non-numerical factors need to be considered. It could be that in the past whenever it has precipitated I have had great fun standing outside for hours enjoying every moment that the delicate raindrops come into contact with my skin; in this case the scientific number could cause me to change my course of action so that I have time to go outside later in the day. However, it could be that I cannot stand the rain; in this case the scientific number will lead to other possible actions, such as taking my umbrella with me when I leave the house, or changing my plans so that I can stay at home all day and don't have to venture outside.

A number is just a number. A measurement is just a measurement. One cannot move straight from a number or a measurement to a conclusion concerning an appropriate course of action. In the case of global warming, one cannot move straight from the scientific measurements and numbers relating to the phenomenon to a conclusion concerning the appropriate human response. Strictly speaking, a measurement or number cannot even reveal that there is a problem. The fact there is a 95% chance of precipitation is not a problem to me if I enjoy precipitation or if I dislike it but intend to stay inside all day. Similarly, the fact that human activities have resulted in increasing atmospheric greenhouse gas concentrations is only a problem if one adds to the measurement the assumption that the future survival and wellbeing of the human species and the other life-forms of the Earth is important. However, I will assume that you agree with me that the future survival and wellbeing of human and non-human life on Earth is important. This means that we can fruitfully speak of there being a scientifically-revealed problem; it is just that the nature of the solution to the problem is not automatically generated by the scientific measurements and numbers.

Finding the appropriate solution to the problem revealed by the science of global warming requires a consideration of a whole range of non-scientific non-numerical factors. What exactly are these factors? These factors are philosophical in nature and jointly constitute the philosophy of global warming. These factors will be explored throughout the rest of this book and include the following:

- The fact that there are two types of global warming (non-human-induced and human-induced) and the relationship between them.

- The question of whether the evolution of human culture has a particular trajectory, a trajectory which includes the environmental crisis and human-induced global warming as essential parts.

- The nature of the relationship between the human species and non-human life on Earth.

- The cosmic, and planetary, significance of technology.

- The extent to which humans, individually and collectively, have freedom to evolve differently to the way that they actually evolve.

- The nature of the Universe, the Solar System, and the Earth; the way that they evolve through time and the way that they 'interact' with each other.

- The relationship between technology, spirituality and the environmental crisis.

- The diverse aspects of the environmental crisis – climate change, sustainability, global warming, biodiversity loss, resource depletion and care for the environment.

In the *Introduction* we saw that there are two ways in which humans might be able to stop carbon dioxide concentrations in the atmosphere from rising too much:

Path 1: Humans stop emitting, or radically reduce emissions of, carbon into the atmosphere.

Path 2: Humans use technology to regulate the amount of carbon dioxide in the atmosphere.

The appropriate choice of path requires a consideration of the range of philosophical factors outlined above. If one does not fully engage with these factors, if people *en masse* lazily opt for Path 1, then the consequences will be adverse if it turns out to be the wrong path. In the next chapter we will take a step backwards and consider the nature of the two paths themselves. These two paths are a particular expression of two 'wider' paths, two general views of the appropriate relationship between the human species and the non-human Earth. In the next chapter the two specific paths outlined above are placed within the context of their 'wider' paths. In the rest of the book we will consider at length the factors outlined above and in so doing our hope will be that the appropriate path will reveal itself.

Chapter 2

The Two Paths Facing Humanity

We are living at an exceptionally important time, a unique moment in the evolution of the planet. Humanity stands at a crossroads and the future is uncertain. There are two possible paths that we can take. One path leads to a glorious and wonderful future, the other leads to death and destruction. Which path will we tread?

Many people know that we are at a crossroads; they see the two paths stretching into the future. They know that one path leads to a glorious and wonderful future and that the other leads to death and destruction. However, the paths are not transparently labelled; they are not labelled 'path to a wonderful future' and 'path to destruction'; if they were the choice of path would be extremely easy. Is it obvious which path leads to which destination? Many people believe that it is. A great many people are certain that one particular path is the path to a glorious and wonderful future; these people are passionately attempting to convince us to walk along this path. This is why I am concerned about the future. We should remember that all that glistens is not gold; what seems obvious at first sight can be wrong, fatally wrong. It is clear to me that the path these people seek actually leads to death and destruction.

What is one to do when one sees good-intentioned people seeking to shepherd humanity along a path which leads to the death and destruction not only of the human species, but of all life on Earth?

One could just sit back and do nothing. Alternatively, one can seek to illuminate the true nature of the two paths so that destruction can be avoided and the path to a glorious future can be trodden. Let us seek this illumination.

The Two Paths

What is the nature of these two paths? The two paths represent different ways in which humanity can interact with the Earth in the future. The choice of path is a very serious affair. There is little more serious than the issue of whether one's species goes extinct and one's planetary home becomes lifeless. Let us choose our path with extreme care.

The two paths are characterised by the amount of involvement humanity has with the rest of the Earth. The first path involves minimalizing involvement. This path has many aspects such as restricting the size of the human population, restricting the human appropriation of the Earth's resources and restricting the deployment of human technology. Those who urge us to tread this path believe that human involvement with the Earth is already too high. These people believe that a high level of future human involvement is a negative thing, both for the human species itself and for the non-human life-forms which we share the planet with. This view is underpinned by the belief that the optimum state of the Earth is one in which human involvement is minimised because humans are fundamentally destructive. Through their greed, their appropriation of the Earth's resources, their technology, humans are seen as a danger to both themselves and to the non-human life-forms of the

Earth. Let us refer to this path as the 'minimalizing involvement' path. The extreme advocates of this view seek absolute minimization – the voluntary extinction of the human species for the good of the Earth. However, the 'minimalizing involvement' path more typically involves calls for restrictions, and a general pulling back of human involvement with the 'non-human', rather than absolute minimization. I refer to this path as 'minimalizing involvement' because minimization is the underpinning ideal. In reality, very few advocates of this path think that involvement should actually be absolutely minimized through the voluntary self-extinction of the human species. All advocates of the 'minimalizing involvement' path believe that significantly reducing human involvement with the Earth would be a good thing. In the specific realm of global warming this 'wide' path is enshrined in the narrower Path 1 which we identified in the *Introduction* and *Chapter One*.

The second path is obviously very different; it involves much more human involvement with the Earth. However, it does not involve 'maximizing involvement'. Absolute maximization would entail humans actively and intentionally utilising, manipulating and moulding every single part of the Earth. Whilst there are those who advocate absolute minimization, I am not aware of anyone who advocates absolute maximization. Indeed, the notion barely even makes any sense (humans would need to be moulding and manipulating every life-form, every ocean, every part of every ocean, volcanoes, the inner core of the planet, etc.). The second path does not entail either absolute maximization or any weaker type of maximization; maximization is not an ideal underpinning the view. The second path involves significantly increasing human involvement with the Earth, but this is a far cry from maximization.

There are two reasons why one might advocate treading this path. Firstly, one might tread this path with regret because one believes that humanity has perturbed the biogeochemical cycles of the Earth to such an extent that our future survival depends on deepening our involvement. For instance, one might love to tread the first 'minimalizing involvement' path, but one believes that human-induced global warming is a serious threat to the future survival of the human species and that it can only be dealt with by geoengineering the temperature of the atmosphere; so, one decides to tread the second path with regret; given the reality of the situation we face, this is the best path to take. Secondly, one might joyously skip and jump along the second path. In other words, one believes that significantly increasing human involvement with the Earth is actually a good thing, a positive event which benefits not only the human species but also the totality that is life on Earth. Let us refer to this path as the 'increasing involvement' path. The advocates of this path believe that significantly increasing human involvement with the Earth would be a good thing – either good solely for humanity or good for the totality of life on Earth. In the specific realm of global warming this 'wide' path is enshrined in the narrower Path 2 which we identified in the *Introduction* and *Chapter One*.

Which of the two paths should we tread?

The Crossroads

Before one can see the true nature of the two paths one needs to clearly see the journey that has led to the crossroads. If one is at the crossroads and is unable to see the entire journey which has led to the crossroads – the journey of the Earth and the journey of the Solar System – then it is unlikely that one will make a good choice of path: 'minimizing involvement' or 'increasing involvement'. One could get lucky and just happen to pick the right path, the path to a wonderful future rather than the path to death and destruction. However, if one cannot see the entire journey which has led to the crossroads then one will not have the tools with which to make a properly informed decision. This means that one could easily select the wrong path, and given the seriousness of the choice this would be a terrible outcome. Furthermore, and worryingly, given the nature of the two paths, it is likely that one's lack of backwards vision will result in one choosing what I believe to be the wrong path. As we saw in the *Introduction* there is a simplistic instinctive response that Path 1 – the 'minimizing involvement' path – is the path to a glorious and wonderful future. A little knowledge can clearly be a dangerous thing. Let us expand our knowledge; let us move beyond the simplistic instinctive response; let us consider the journey that led to the crossroads.

Understanding the past will help us to select the right path in the present.

Chapter 3

Two Types of Global Warming

Do you know that there are two different types of global warming? A great many people seem to believe that there is only one type of global warming; they believe that global warming is caused by human activities and that this is the only type of global warming. The two types of global warming that exist are human-induced global warming and non-human-induced global warming. It is important to distinguish these two types of global warming from other short-term factors which affect the temperature of the atmosphere.

Most people realise that there are small variations in the temperature of the Earth's atmosphere that are caused by non-human factors. Some of these non-human factors are: a slight change in incoming solar radiation due to the sunspot cycle of the Sun; volcanic activity which temporarily changes the composition of the atmosphere through increasing the amount of aerosols; and, a switch in the El Niño Oscillation. At a particular moment in time any of these factors could be a non-human cause of a slightly warmer atmosphere; these factors therefore complicate the figures and models of the science of global warming. These short-term factors which can increase the atmospheric temperature shouldn't be thought of as a type of global warming. They are simply cyclical variations and short-term phenomena which naturally get reversed (the short-term warming is automatically followed by an opposing cooling).

To talk of global warming is to talk of a *non-cyclical one-way process*. The sunspot cycle is a cyclical cycle; the level of incoming solar radiation to the Earth varies in accordance with the cycle. This means that the forces for warming and cooling of the temperature of the Earth's atmosphere correspondingly vary over the cycle. Over time the net effect on the Earth's atmosphere is neutral. The same is true for volcanic activity and the El Niño Oscillation. So, to say that there are two different types of global warming is to say that there are two non-cyclical one-way forces which are exerting upwards pressure on the temperature of the Earth's atmosphere.

If we are to choose the right path (Path 1 or Path 2) one of the most essential things to realise is that two different types of global warming exist – human-induced global warming and non-human-induced global warming. If the choice of path is made based on the erroneous assumption that there is only one type of global warming, then a bad selection is much more likely; the choice will be made on the basis of very limited information. Given that the future of the human species and that of a plethora of non-human life-forms is at stake, the choice of path should not be made by people who are blinded to the bigger reality.

We have already considered the phenomenon of human-induced global warming in the *Introduction*. The result of human activities on the Earth has been to create a non-cyclical one-way force for global warming. We can imagine a hypothetical scenario in which human activities on the Earth are cyclical in relation to global warming; fossil fuels would be released from underground storage in one century, and the following century they would be buried again; rainforests would be destroyed in one century, and the following century they would be replaced. This hypothetical cyclical scenario can be compared to reality, which is that human-induced global warming is a non-cyclical one-way force. Human activities over time have continuously produced a stronger and stronger global warming force. From thousands of years ago to the present, the rainforests of the planet have been progressively destroyed and replaced with agriculture and urban sprawl. The history of fossil fuel use is one of increasing transfer from underground storage to the biogeochemical cycles of the Earth (no carbon has, so far, been reburied). As we stand at the crossroads, we can see that these human activities have given rise to the one-way force that is human-induced global warming.

To be clear, there have been, and are, particular humans and communities who live in a sustainable, global-warming neutral, way; what we are talking about is the net effect of the total human impact on the planet. This net effect is a non-cyclical one-way force for global warming. It is also important to realise that it is possible for some non-cyclical one-way forces to cease, to become a non-force. To talk of human-induced global warming as a non-cyclical one-way force is not to make predictions about the future; it is simply to recognise the reality of the past and the present. It is to say that this force has existed throughout human history and that it still exists today. If human technology is successfully deployed to actively regulate the atmospheric temperature, then the force will no longer exist; the net effect of human activities would then be a non-cyclical one-way force for the sustainability of the atmospheric temperature.

Let us now consider the second type of global warming – non-human-induced global warming. When we consider the history of the Earth, from its formation to the present, then the overwhelmingly dominant force exerting upwards pressure on the temperature of the Earth's atmosphere is non-human-induced global warming. The source of non-human-induced global warming is the increasing amount of solar radiation which reaches the Earth from the Sun. Since life arose on the Earth the amount of solar radiation reaching the Earth from the Sun has increased by a massive 25%. As we have seen, there are short-term cyclical fluctuations in solar radiation which correspond to the sunspot cycle, but these are superimposed onto a continuous long-term one-way upwards process, the forever increasing amount of solar radiation being propelled towards the Earth. This one-way process gives rise to the type of global warming that is non-human-induced global warming.

Now is a good time to stress the crucial difference between the existence of a 'force for global warming' and 'the actuality of global warming'. To talk of a type of global warming is simply to say that there is a force for global warming, whether or not this force has yet manifested itself into the actuality of global warming. What is important are the forces for global warming which are coming into actuality rather than whether the effects of these forces have yet to be unleashed. Imagine that someone is holding an elastic band next to your face by letting it hang of its own accord from their finger, in this case there is no force waiting to be unleashed. However, when they put their other hand through the loop and start drawing the elastic band further and further backwards an increasing force is being generated which is yet to be unleashed. One can say that there is a non-cyclical one-way force for 'face stinging' being generated, although there is no 'actuality of face stinging'. This same process applies in the global warming realm; it is the existence of forces for global warming which is more important than whether these forces have so far manifested actual global warming.

Clearly, real sustainability of the atmospheric temperature either requires the absence of any forces for global warming, or the continuous long-term sustainable counterbalancing of a force for global warming with a force for global cooling.

The former possibility for real sustainability is not even an option because non-human-induced global warming is a force that will never cease. The latter counterbalancing option is possible, and this counterbalancing has actually been happening on the Earth for millions of years through planetary homeostatic regulation. However, as we will see, the long-term sustainability of this counterbalancing ultimately requires the deployment of technology as a counterbalancing force for global cooling. Let us now return to our consideration of the type of global warming that is non-human-induced global warming; we can then explore how this force has been offset by planetary homeostatic regulation.

The 25% increase in solar radiation which has reached the Earth from the Sun is the dominant type of global warming on the Earth. The *force* for global warming which this increase has generated are immense; yet this increase hasn't actually increased the temperature of the Earth's atmosphere; in other words, the 'force' has yet to manifest as an 'actuality'. It is essential to understand why this is so. This is essential because such an understanding reveals a picture of the Earth as a dynamically unfolding self-regulating whole, and human activities need to be seen as part of this unfolding process. The short answer as to why the 'force' has yet to manifest an 'actuality' is that there has been an equally opposing force (planetary homeostatic regulation) which has offset this immense force for global warming. This is the short answer; a much longer answer is clearly required.

So, here we are standing at the crossroads with the two paths laid out in front of us. If we are to choose the right path we clearly need to understand the past dynamics that have been operating in the realm of non-human-induced global warming, for, these dynamics are still in operation as you read these words. The key dynamic we need to explore is why a 25% increase in incoming solar radiation to the Earth – a massive force for non-human-induced global warming – hasn't yet translated into the actuality of global warming.

Let us start with the Sun. The main factor determining the temperature of the Earth's atmosphere is the amount of solar radiation that emanates from the Sun and reaches the Earth. This amount varies through time, increases, as the Sun ages. Since life arose on Earth (3,500 million years ago) to the present day incoming solar radiation has increased by 25%. In the future this amount will continue to increase as the Sun continues to age and ultimately expire. As Sir James Lovelock puts it:

> We may at first think that there is nothing particularly odd about this picture of a stable climate over the past three and a half eons [3,500 million years]... Yet it is odd, and for this reason: our sun, being a typical star, has evolved according to a

standard and well established pattern. A consequence of this is that during the three and a half aeons of life's existence on the Earth, the sun's output of energy will have increased by twenty-five per cent.

(*Gaia: A New Look at Life on Earth*, OUP, 2000, p. 18)

Next we need to realise that life-forms need particular conditions in order to survive and that the temperature of their surroundings is a particularly important condition. You may well have heard the phrase the 'Goldilocks Zone' which refers to a location which is neither too far from, nor too close to, the Sun. Planets within the 'Goldilocks Zone' are likely to have conditions which might be suitable for life – notably an atmospheric temperature which is neither too cold for life nor too hot for life. There are some simple life-forms, such as bacteria, which can exist in relatively extreme temperatures. However, the vast majority of life-forms ('complex life') require a very narrow temperature range in order to survive. Since life arose on the Earth the average temperature of the atmosphere (the Global Mean Surface Temperature/GMST) has always been between 10°C and 20°C, and this is precisely the range that complex life-forms need in order to survive and thrive.

So, we have two facts. Firstly, incoming solar radiation to the Earth increases over time. Secondly, in order to survive and thrive, life on Earth requires a relatively narrow GMST of between 10°C and 20°C. From this follows a seemingly straightforward conclusion: *if the amount of incoming solar radiation that reaches the Earth is either too low or too high then the temperature range which life requires will be breached.* When life arose on the Earth the amount of solar radiation reaching the planet was at the low end of what life requires. Since then there has been a 25% increase in solar radiation reaching the Earth, a massive force for global warming, yet the temperature of the atmosphere has not increased; throughout the 3,500 million year period it has fluctuated within the 10°C to 20°C range. How can this be? If the Earth was not responding to this increase in incoming solar radiation the upper 20°C ceiling would have been breached a very long time ago. As Sir James Lovelock explains:

> right from the beginning of life, around three and a half aeons ago, the Earth's mean surface temperature has never varied by more than a few degrees from its current levels. It has never been too hot or cold for life to survive on our planet, in spite of drastic changes in the composition of the early atmosphere and variations in the sun's output of energy.
>
> (*Gaia: A New Look at Life on Earth*, OUP, 2000, p. 48)

> If our planetary temperature depended only on the abiological constraints set by the sun's output and the heat balance of the Earth's atmosphere and surface, then... all life would have been eliminated.
>
> (*Gaia: A New Look at Life on Earth*, OUP, 2000, p. 20)

In other words, life on Earth has counterbalanced the increasing solar radiation (a force for global warming) by responding with an offsetting force, a force for global cooling, a force to maintain the temperature of the atmosphere so that complex life-forms can continue to exist, continue to thrive, and continue to evolve. What is the nature of this counterbalancing force that the Earth has deployed? One of the main ways that the Earth has regulated the temperature of its atmosphere is by removing greenhouse gases from its atmosphere and storing them under the surface of the Earth. Less greenhouse gases in the atmosphere means a smaller 'greenhouse effect' and a lower GMST; this is thus a powerful force for global cooling to offset the force for global warming that is increasing incoming solar radiation. One way of putting this is to say that the Earth has been homeostatically regulating itself in order to maintain the conditions that life needs in order to survive and thrive.

Let me reiterate and reword this for the sake of clarity. It has only been very recently that scientists have discovered that the level of greenhouse gases in the atmosphere plays a pivotal role in determining the temperature of the atmosphere, yet for millions of years the Earth has been actively removing greenhouse gases from its atmosphere with the result that its temperature could be maintained so that complex life-forms could exist on the Earth. The force for global warming, the 25% increase in the output of the Sun, has been offset by an opposing force deployed by the Earth, the removal of greenhouse gases from the atmosphere to maintain the atmospheric temperature. Before humans evolved on the Earth it was not a harmonious peaceful place; there had been a great battle on-going for millennia upon millennia, epoch upon epoch. Two great opposing forces were locking horns against each other. The Sun, sending out ever-increasing solar radiation; the Earth, retaliating by removing greenhouse gases from its atmosphere. The result is an on-going battle for the atmosphere, an on-going battle for the survival of life on Earth.

This battle appears to be between enemies of unequal power; an epic 'David versus Goliath' confrontation. The Sun is Goliath; it is unrepentant in its ability to send forever increasing amounts of solar energy to the Earth. It is well-known that the Sun is going to continue to get hotter and hotter; it is going to keep on sending more and more solar radiation to the Earth, until in the distant future it expires. However, our concern is not with the distant future; it is with the past, the present and the immediate future. The Earth is David; it is an inferior opponent, its weapons are weaker, it is smaller and more fragile; yet it is seemingly determined to be victorious, it is still resisting the increasing solar radiation today despite the increasing difficulty of its task. This is a key point to fully appreciate; the expiration of the Sun is a very long way in the future, but the on-going battle we have been exploring means that life on Earth faces the threat of extinction from the increasing output of solar radiation from the Sun in the very near future. A tipping point will be reached when the ability of the Earth to resist, to counterbalance, the increasing solar radiation, will end. The signs are that this time is fast approaching. As Sir James Lovelock puts it:

> [Gaia] is old and has not very long to live. As the sun grows ever hotter it will, in Gaia's terms, soon become too hot for animals and plants and many of the microbial forms of life.
>
> (*The Revenge of Gaia*, Penguin Books Ltd., 2006, p. 46)

only for a brief period in the Earth's history was the sun's warmth ideal for life, and that was about two billion years ago. Before this it was too cold for comfort and afterwards it has progressively grown too hot... The sun is already too hot for comfort.

(*The Revenge of Gaia*, Penguin Books Ltd., 2006, p. 44-5)

We have seen that the Earth has been defending itself by removing carbon from its atmosphere and burying it under the surface of the Earth. The more carbon that is buried the greater is the force for global cooling. This has been the prime weapon deployed by the Earth for millions of years in order to offset the increasing incoming solar radiation from the Sun. It has been a successful way of keeping the planet habitable for complex life-forms; indeed, life has positively thrived on the Earth.

The reason that the Earth-Sun confrontation is a 'David versus Goliath' confrontation is that whilst the Sun has unfaltering reserves of solar radiation, the Earth only has a limited amount of carbon dioxide in its atmosphere. In other words, moving carbon from the atmosphere to underground storage, via vegetation, is a process that becomes increasingly difficult over time. According to one prominent line of thought, one that seems to me to be correct, the fact that the Earth entered into a period of transition between glacials and interglacials, in accordance with Milankovich Cycles, is a sign that the Earth has been struggling to maintain the temperature of its atmosphere for quite some time. In other words, these transitions are a sign of planetary sickness, a sign that the Earth's ability to effectively draw down carbon dioxide from its atmosphere and safely store the carbon underground is severely weakening. These transitions are a sign that life is in peril from non-human-induced global warming. As Sir James Lovelock puts it:

> The brief interglacials, like now, are, I think, examples of temporary failures of ice-age regulation.
>
> (*The Revenge of Gaia*, Penguin Books Ltd., 2006, p. 45)

With this evolutionary background in mind, let us consider the evolution and activities of the human species; let us see how human-induced global warming fits into the on-going battle we have been considering. The Earth has managed to continue the battle with its more powerful opponent for long enough to allow a species as complex as the human species to evolve. The human species is the technological part of the Earth; the bringer forth of technology. The human species has, through its technological ability, removed massive amounts of the carbon that was previously stored underground by the Earth. This carbon was stored underground by the Earth in order to keep the temperature of the atmosphere constant despite the increasing incoming solar radiation. Now humans have released much of this carbon back into the biogeochemical cycles of land-ocean-atmosphere. This seemingly has potentially catastrophic consequences. This release is the prime source of the force that is human-induced global warming. The underground storage of carbon was a force for global cooling,

Two Types of Global Warming

its release by humans is a force for global warming; a force for human-induced global warming.

The question of the appropriate human response to global warming needs to be seen in the context of this evolutionary background: the existence of two types of global warming and the evolutionary battle for the atmosphere between the Earth and the Sun which long precedes human existence. In the context of this on-going battle it seems entirely plausible to me to believe that the Earth brought forth the human species precisely when it needed a technological species. The human=technological species has evolved on the Earth at precisely the moment in the evolution of the Earth that the Earth needs a technological species if life on Earth is to continue to survive and thrive. The Earth was inevitably losing its battle with the Sun, and the continued existence of complex life on Earth required something new, it required a technological species which could technologically control the temperature of the atmosphere so as to keep it favourable for life on Earth. I find it hard to believe that it is simply a coincidence that the Earth brought forth a technological species at just the moment in its evolutionary unfolding that it required a technological species if it was to continue to maintain the atmospheric conditions that are required to support complex life.

So, whilst the human release of carbon *seemingly* has potentially catastrophic consequences, in this larger context the human development of technology has potentially wonderful and life-saving consequences. The human development of technology has simultaneously exacerbated the pre-existing pre-human problems that the Earth was having homeostatically regulating its atmosphere (by being utilised to release carbon from underground storage), and provided the solution to this pre-existing problem (the need for technological regulation of the GMST).

We have been considering the two paths that face us at the crossroads:

> **Path 1:** Humans stop emitting, or radically reduce emissions of, carbon into the atmosphere.
>
> **Path 2:** Humans use technology to regulate the amount of carbon dioxide in the atmosphere.

Our considerations of non-human-induced global warming in this chapter lend strong support to the idea that Path 2 is the appropriate path. More than this, they suggest that a more profound path is required for the future existence of both human and non-human complex life on Earth:

> **Path 3:** Humans use a whole range of technologies to regulate the *temperature* of the atmosphere.

As we have seen, there comes a point when simply regulating the amount of greenhouse gases becomes ineffectual; greenhouse gases have become minimised but incoming solar radiation continues to increase. There is currently a delicate interplay of forces occurring on the Earth. The non-human ability of the Earth to offset the increasing incoming solar radiation has been weakening because of the reduced atmospheric greenhouse gas concentrations which could potentially be drawn down to underground storage. However, in the short-term humans have released a massive amount of carbon from its underground storage areas; this means that if it is stored underground once again then this would be a short-term force for global cooling.

It is important to realise that such re-storage does not happen overnight (in the absence of technology), it takes an exceptionally long time; furthermore, such re-storage would only be a transitory 'sticking plaster' solution, and it would ultimately be ineffectual given the broader unfolding situation. The underlying problem (the forever increasing incoming solar radiation) would remain. This is why a whole range of technologies are required. To be clear, if the only type of global warming was human-induced global warming, then simply reversing what humans have done, drawing back down the carbon which we have released from underground storage, would be the end of the story. But this isn't the end of the story; it is but the beginning. We can use technology to draw down carbon from the atmosphere and return it to underground storage and this will buy us some time. However, the fundamental type of global warming is non-human-induced global warming and dealing with this requires more than technologies to regulate greenhouse gas concentrations in the atmosphere; it ultimately requires technologies to block the incoming solar radiation from reaching the atmosphere of the Earth in the first place.

To achieve this objective, I believe, is the reason why the human species was brought forth into existence, and it is the reason why we are the most important species on the planet. You might not be convinced of this yet, but there is much more to be said.

Chapter 4

The History of Our Solar System

The purpose of this book is to consider the appropriate human response to global warming. I should remind you that I am using the term global warming to refer to the two non-cyclical one-way forces that are exerting upwards pressure on the temperature of the Earth's atmosphere. In other words, the term global warming refers to *both* human-induced global warming and non-human-induced global warming.

What does the appropriate human response to global warming have to do with the history of the Solar System? We need the broader perspective of the Solar System if we are to get a fuller understanding of the situation that we are currently in; it is the understanding of a situation that determines how one responds to that situation. In other words, if one broadens the horizon of the factors that one considers then one can come to a different conclusion to when one only has a narrow horizon. Another way of putting this is to say that there are two ways of approaching a problem. Firstly, one can seek to remove as many factors as possible and consider only those factors which seem to one to be of pivotal importance. Secondly, one can seek to appreciate a whole host of interconnected factors so that the problem can be seen in a larger context. This second approach is our approach.

The Solar System is an evolving entity which can fruitfully be compared to a human body. The human body has a heart which is the lifeblood of the body; the heart is surrounded by various other organs/limbs/body-parts. Some of these body-parts are close to the heart, others such as the feet and the ears are further away. The human body evolves, ages and dies. The Solar System is very similar. It has the Sun which is its lifeblood, its central heart which powers the Solar System with its flows of solar radiation. The Sun is surrounded by various parts, particularly 'nine' planets, some of which are fairly close to it and some of which are much further away (just like the body-parts surrounding the heart in the human body). The Sun and the Solar System evolve, age and die, just like the human body.

It is easy to think of the human body as an interconnected whole, in which what is going on in one part is of importance for the other parts. For example, thinking about the answers to a set of exam questions is likely to be more difficult if one has excruciating pain in one's legs. This feat of appreciation – appreciating the interconnectedness of various parts – is less easily achieved for the Solar System. To see the interconnections between the celestial bodies, and the situations and problems on the Earth, is much harder.

I would like you to try and envision the Solar System as an interconnected whole. The Sun is the heart of the Solar System; it is the radiator of the energies which can bring forth life in the Solar System; wonderful, precious life. The planets that are closest to the Sun are too close, too hot, for life to survive and thrive. The planets at the extremities of the Solar System are too far away, too cold, for life to survive and thrive. The Earth is the part of the Solar System where life is meant to thrive, just as the lungs of the human body are the part which enables a human to breath. There is only one part of the human body which can enable a human to breath; without the lungs there is no breath. There is only one part of the Solar System which can enable life to thrive; without the Earth there is no thriving. The Earth is the womb of Solar-Systic Life.

The Earth has sustained life for most of its history. Life has evolved from simple beginnings to highly complex animals. Complex animals have evolved culture, and human cultural evolution has similarly evolved from simple beginnings to globalised technological society. Taking the Earth as a whole there is a continual evolution from the simple and unconnected, to the complex and interconnected. We are a part of this great unfolding and evolving process. You might have been told that the evolution of life-forms is solely down to natural selection and that humans evolved as a fluke. What a limited understanding! The evolutionary processes of life and of human culture are underpinned by cosmic forces. The Earth is the womb of Solar-Systic Life, but that life is propelled forward by the entire body, the entire Solar System. The development of a baby in the womb of its mother is not isolated from the actions of its mother. Similarly, the development of life on Earth is not isolated from the movements of the Solar System (as has been known by astrologers for a very long time). As the planets swirl through their orbits, life on Earth gradually evolves from simple beginnings to globalised technological society. Swirl, evolve; swirl, evolve; swirl, evolve. From the beginnings of the Solar System, the evolution of life on Earth, the evolution of the human species, and cultural evolution to the globalised technological society of today, was effectively 'programmed in'. The formation of the Solar System can be compared to the programming of a 'Sat Nav' in a car; once it was formed, both the destination and the path to that destination were already largely determined.

Where does global warming fit into this picture? We have already explored the two different types of global warming. A fundamental part of the evolving Solar System is that as the Sun ages it sends out more solar radiation; the result is the global warming of the Solar System. This means that over time, as the Solar System ages, the planets closest to the Sun become increasingly hotter and even more inhospitable to life; whereas the planets at the extremities of the Solar System also receive increasingly more solar radiation, yet they still remain too cold for life to survive and thrive. There is only a limited temporal window in the unfolding of the Solar System within which non-human-induced global warming interacts with planetary location in order to enable the emergence, survival and thriving of life. In other words, the womb of Solar-Systic life comes into existence and then it either gives rise to a successful birth, or it goes through a painful abortion.

We have already explored how the Earth stored fossil fuels underground so that it could offset the global warming of the Solar System and thereby maintain the Earth as the womb of thriving life. We have also explored how humans have released a massive amount of these fossil fuels. It is best to think of this original process of storage as an instinctive, almost automatic, part of the unfolding ageing Solar System. In other words, the process didn't require conscious knowledge or foresight of what was required to keep the temperature of the Earth's atmosphere hospitable for planetary life. This process is just what happens on a thriving life-bearing planet which is located in an ageing Solar System.

Let us return to the topic of the human release of these stored fossil fuels. This release is the dominant cause of human-induced global warming. If the evolution of life and culture to globalised technological society on the Earth was 'programmed in' at the formation of the Solar System, then this means that human-induced global warming was also 'programmed in'. This is because the attainment of globalised technological society requires the appropriation of the Earth's resources on a massive scale; in particular it requires the use of the fossil fuels that were to be discovered by humans under the surface of the Earth. In other words, we can see human-induced global warming and the wider environmental crisis as part of the planetary and cosmic unfolding process. In the same way that the initial burying of the carbon was a necessary activity for maintaining an atmospheric temperature conducive for thriving life, the human release of this carbon can be seen as part of this same unfolding process. At the earlier stage in the development of the Solar System the appropriate activity – the instinctual, automatic, unfolding activity – was for carbon to be buried under the Earth's surface. At this later stage of development the appropriate activity – the instinctual, automatic, unfolding activity – is the human release of the carbon.

It is clear to me that life's appropriation and modification of the Earth's resources, via the human species, is a necessity if the Earth is to continue as the womb of Solar-Systic Life. The womb needs increasing protection, and it needs to protect itself through marshalling its resources into the forms we call 'technology'. Technology can help to protect life on Earth in many ways, but the most pressing need is to keep the atmospheric temperature within a range that is suitable for life to flourish in the face of the increasing output of the Sun. The bringing forth of technology brings forth both the solution to this problem as well as a more pressing need for its implementation. This is because the development of technology to combat non-human-induced global warming simultaneously brings forth human-induced global warming. Without technology there will be a painful abortion.

If this is making sense to you then you will understand what I mean when I say that human-induced global warming is a sign that the Earth is flourishing. Human-induced global warming is a sign that life on Earth is flourishing. Human-induced global warming is a sign that the unfolding of the Solar System has reached a crucial stage. Human-induced global warming is a sign that the womb of Solar-Systic life is approaching maturation.

We should be celebrating, yet everywhere there seems to be panic and despair.

Chapter 5

What is Life?

My aim is to convince you that the technological regulation of the Earth's atmosphere by the human species is a positive event which should be fully and joyously embraced. Furthermore, I seek to convince you that it is a positive event primarily for the life that has arisen on the Earth. As the human species is a part of that life this means that it is also a positive event for the human species. Furthermore, as the Earth is the womb of life in the Solar System, it is also a positive event for the Solar System. In order to convince you of this I need to be very clear about what it means to talk of life and also what it means to talk of the human species as part of life. In this chapter I will explore various terms relating to life; in the following chapter I will explore what it means to talk of the human species.

So, what is life? I am sure that you probably think that you have a pretty good understanding of which parts of the Earth are living and which are non-living. When you are walking through a park you encounter easily identifiable living and non-living things. The humans you see walking in the park are living, so are the dogs that are running in the park, as are the birds in the trees, the trees, the squirrels, the worms, the cats, the shrubs, the grass and the flowers. In contrast, the wooden fence is non-living, so is the lead which connects dog to owner, and the benches, the rubbish bins, the street lamps and the concrete paths. This kind of division between the living and the non-living is surely correct.

However, there are lengthy, on-going and unresolved debates concerning what exactly life is. There are seemingly borderline cases where it is questionable whether an entity is living or not. However, there is a deeper epistemological issue arising from the possibility that the division between the living and the non-living is a division that humans project onto the world, rather than a division in the universe itself. Thinking along these lines results in some people denying that there is actually a division between the living and the non-living; these people typically believe that the entire universe is living. I don't think that we should be tempted by this line of thought. It seems obvious to me, and surely to you as well, that there is a real division in the universe between the living and the non-living; a division which exists independently of human perception, thought and classification.

Let us consider the living/non-living division from the perspective of the evolving Solar System. If we go back far enough in time the Solar System was in a wholly lifeless state; there was no division between the living and the non-living because the Solar System was wholly constituted by the non-living. Then, when the conditions were right, the non-living brought forth the living; the evolving Solar System brought forth life. Life might have arisen on non-Earth locations, locations such as Mars, where the conditions enabled life to arise, but where the conditions were not suitable for life to thrive; on such locations a temporary chasm opened and then closed. These parts of the Solar System would have moved from A) Lifeless to B) Living/Non-living and on/back to C) Lifeless.

On the Earth the chasm opened, life was brought forth, and as the conditions were suitable for life to thrive, the chasm has remained open; there is a division between the living and the non-living. In the future, if we extend far enough in time, eventually the chasm will close; when the Sun expires, if not before, the Earth will again become a lifeless part of the universe. This way of thinking about life is useful because it helps one to grasp the idea that life is a single entity. Life is that part of the Earth which is not the non-living. Life is that part of the Earth which wouldn't exist if the Earth fell back to a state of lifelessness.

So, the Earth can be divided sharply into two parts – the living and the non-living. I will generally use one of two phrases to refer to the living part: 'life on Earth' or 'planetary life'. These phrases are used to encompass everything on the Earth that is alive. At a broader level, I typically use the phrase 'life' to refer to life wherever it exists in the universe. At a narrower level, I divide 'life on Earth' into two parts. It is useful to think of 'life on Earth' as encompassing 'complex life' and 'non-complex life'. 'Non-complex life' is a term which refers to extremely simple life-forms: bacteria, single-celled life-forms and other microbes. These life-forms can exist in extreme conditions and can therefore exist when a planet is not thriving. The division is important because when life is firmly established on a planet it evolves 'complex life' which is a sign that life is thriving. If you find life-forms on a planet such as cats, dogs, bats, dinosaurs, trees, elephants, dolphins, roses, carrots, and birds, then life is thriving on that planet.

Thinking of life in this way, as a single entity, or as a single entity with two parts, is not how most people typically think about life. Most people typically think about life in terms of individual organisms. They think of life as a term which simply refers to an enormous number of individual organisms. In other words, individual organisms are aggregated so as to form life. So, life on Earth at a particular moment in time could be thought of as being constituted by 5678 dolphins + 233000 ladybirds + 60000000 humans + 70000000 trees, and so on. If you can give up this individualistic and aggregative way of thinking about life, and move to a visualistic and holistic way of thinking, then this will aid your understanding. Try and visualise life on Earth as a single interconnected entity which spans the planet; the birds in the sky, the worms in the earth, the fish in the sea, the humans in the cinema – all parts of a single entity. This entity is life on Earth/planetary life. Planetary life isn't simply a collection of individual life-forms; it is a single entity that is constituted out of what we call individual life-forms. If this distinction isn't clear to you it may help to think of your body; you probably think that your body isn't simply a collection of individual parts, rather, it is a single entity which is constituted out of 'individual' parts.

When I talk about the interests of life on Earth I am talking about the interests of this single entity. I am not talking about what is in the interests of a dog, or a whale, or a bird, or a human. I am not even talking about what is in the interests of a species of dog, or all dogs, or the human species, or mammals. When I talk about the interests of life on Earth I am talking about what is in the interests of this single entity. This single entity emerged from a state of planetary lifelessness, it created a division between the living and the non-living, and its continued existence is of importance because life is precious. What is in the interests of life on Earth is clearly also in the interests of life, as life on Earth is part of life (it is possible, but unlikely, that life is wholly constituted by life on Earth, rather than life on Earth being a part of life).

Let me briefly return to the epistemological worry concerning the possibility that the division between the living and the non-living is a division that humans project onto the world, rather than an actual division in the universe itself. It is an important part of my philosophical worldview that there is a real division in the universe itself between the living and the non-living, and that the living is a particularly important and precious part of the universe. However, whilst it is obvious to me that there is a distinction in the universe itself between the living and the non-living, it also seems obvious to me that this distinction is a subtle one:

> *One could think of the living part of the universe as a state of 'excitation' or 'exhilaration'.*

There is clearly a distinction between the states of the universe that exist when one is 'exhilarated' compared to the states of the universe that exist when one is 'depressed'; furthermore, a state of 'exhilaration' is surely a better state of the universe than a state of 'depression'. Analogously, a state of living is a better state of the universe than a state of non-living. Whilst there is a clear distinction between 'exhilaration' and 'depression', between living and non-living, these distinctions do not entail the existence of chasms between the parts of the universe that instantiate them. Furthermore, given the subtleness of the distinction, there are good grounds for believing that the evolutionary processes of life are fundamentally similar to the evolutionary processes that occur in the non-living universe. It might sound a bit woolly to say that life is precious because it is a state of universal excitation, but I hope that you can grasp what I am trying to convey.

I have outlined the division that exists between life and non-life; I have also outlined the division that exists within 'life on Earth' between 'complex life' and 'non-complex life'. There is one other division that exists which is of great significance. This is the division that exists within 'complex life' between 'the human species' and 'non-human complex life'. This division is of great significance because the human species is the pinnacle of the evolutionary progression of life on Earth. You might find this assertion to be puzzling, even troubling; it probably contradicts what you have been told by the media, or by academic authorities. It is supposedly intellectually responsible to believe that the human species is just one species amongst many, and that there is no pinnacle or zenith to the evolutionary progression of life on Earth.

Those who peddle this view, curiously, seem to believe that Charles Darwin's theory of natural selection is an adequate explanation of the evolutionary trajectory of planetary life; as if Darwin had the final word on the matter, rather than one of the first. How limited their understanding is! Imagine that the evolution of life on Earth is a tree which grows from a miniscule size to a massive height. Natural selection gives an explanation of the branches and the leaves, but it gives no explanation of the heart of the tree – the trunk. As the trunk of a tree moves upwards towards the sky it provides a direction from which the branches can spread out. As with a tree, so it is with the evolutionary progression of life on Earth. The evolutionary trajectory of life on Earth is determined by the Solar System. The fundamental nature of the cosmos (of which the Earth and the Solar System are parts) in tandem with the swirling of the planets around the Sun (as they enter into particular relations with the Earth) and the concordant ageing of the Solar System, determines the evolutionary trajectory of life on Earth. This is the trunk; natural selection is just the branches. How could anyone seriously believe that the entire evolutionary trajectory of life on Earth is determined wholly by natural selection?! This is like focusing solely on the icing in blissful ignorance of the cake.

We will consider in depth the evolutionary mechanisms which determine the evolutionary trajectory of life on Earth a little later. Our next task is to consider why the human species is the zenith of the evolutionary progression of life on Earth; this requires a consideration of what it means to talk of the 'human species'.

Chapter 6

What is the Human Species?

Having considered what it means to talk of life, we can now consider what it means to talk of the human species as part of planetary life. More than this, we can consider why the human species is the most important part of planetary life. You might not be comfortable with the idea that the human species is the most important part of life on Earth. You might think that all notions of human importance are misplaced because they are wholly self-generated and arise from an egocentric sense of self-importance. You might simply believe that all life-forms are equally important. Additionally, you might believe that numerous parts of life on Earth – dolphins, seagulls, humans, chimpanzees, chipmunks – all believe that they are the most important part of life on Earth. Surely, the argument goes, if all of these parts believe that they are the most important, then, in reality, none of them are the most important. However, this conclusion doesn't follow. It might be true that various parts of life on Earth believe that they are the most important part, but it doesn't follow that in reality none of these parts is the most important part. In order to convince you that the human species is a uniquely important and essential part of planetary life I first need to outline what the human species is.

Recall from the previous chapter that planetary life is a single entity:

> Try and visualise life on Earth as a single interconnected entity which spans the planet; the birds in the sky, the worms in the earth, the fish in the sea, the humans in the cinema – all parts of a single entity. This entity is life on Earth/planetary life. Planetary life isn't simply a collection of individual life-forms; it is a single entity that is constituted out of what we call individual life-forms.

Humans, and all other life-forms that currently exist on the Earth, are part of a single interconnected entity. This means that in a fundamental sense there are no species; there is just a single entity which is instantiated in a multitude of places; species are constructions of the human mind. What humans call 'a bird flying through the sky' is simply a tiny state of excitation that is surrounded by non-excitation. What humans call a 'school of dolphins' is simply a collection of states of excitation which are each immediately surrounded by states of non-excitation. If one had the appropriate viewing equipment one could look at the Earth from a distance and see a wonderful web of interconnected states of excitation, a glowing radiant interconnected web. Part of this web is the bird in the sky; part of this web is the school of dolphins; individual humans are also parts of this web.

I am trying to get you to visualise the life-forms of the Earth as patterns of excitation. It is perhaps helpful to think of life-forms as sparkly glowing entities which are surrounded by darkness. A bird flying through the sky is a particular sparkly glowing pattern which is located some distance above the surface of the Earth; a school of dolphins are sparkly glowing patterns which graciously glide and then leap up and down (as they jump out of the water and then dive back into it); humans sitting in a cinema are comparatively fairly static sparkly glowing patterns.

In a fundamental sense there are no species, there are just individual instantiations of sparkly glowing entities which are part of the interconnected web that is planetary life. There are very varied patterns within this web – the pattern that humans call a bird is very different to the pattern that humans call a dolphin. So, the concept of a species does reflect some aspect of reality. One can think of a species as a particular structure/pattern of sparkly glowing entities (this is the sense in which I will use the term 'species').

> *When I talk of the human species you should keep in mind that I am talking about a particular set of sparkly glowing patterns which exist as part of the total interconnected sparkly glowing pattern that is life on Earth.*

My aim is to convince you that the human species is a sparkly glowing pattern which is fundamentally different from the patterns which are instantiated in all of the non-human life-forms of the Earth. Indeed, we can think of the human species as the highest state of excitation ever to have existed on the Earth. Furthermore, when we consider the mechanisms which have propelled the evolution of life from very simple beginnings to wonderfully complex humans, we can understand this process as being driven by the desire to attain a higher state of excitement. The sparkly glowing patterns radiate ever more brightly as life inevitably moves, over billions of years, from single celled existence to the human species.

What is the Human Species?

Why is the sparkly glowing pattern that is the human species fundamentally different, and superior to, all of the other sparkly glowing patterns on the Earth? The human sparkly glowing pattern is a pattern of exuberance, of joy, of release, of great distance, of great speed, of interconnected pervasiveness. Imagine all of the individual humans currently on the Earth; imagine all of their movements as a sparkly glowing pattern. What a sight this would be to see! Imagine all the movements involved in rush hour in London, New York, Tokyo. Imagine all of the humans inside their cars and imagine the patterns which are formed as these humans whizz around on the motorways; this whizzing spans the globe! Imagine the patterns that are formed by humans as their airplanes glide around in the skies way above the surface of the Earth; what a glorious sparkly pattern circling the globe! Imagine the patterns formed by humans in hot air balloons, helicopters, trains, hand-gliders, submarines; picture humans in skyscrapers, humans bungee jumping, humans skateboarding, humans on the Eurostar, humans on jet skis, on surfboards, on skateboards, on fairground rides, and on canoes. Imagine the patterns formed by the humans in the International Space Station, the patterns formed when humans walked on the moon, the patterns formed by all astronauts. The sparkly glowing patterns are not only whizzing around the Earth at great speed in an immense sense of excitement, of release, they have also broken free from the planet and can be seen extending out to the moon. What joy!

After this visualisation I am sure that you can easily appreciate that the sparkly glowing pattern of the human species is fundamentally different from all of the other sparkly glowing patterns on the Earth. In other words, you can easily appreciate that the human species is different from all of the other life-forms on the Earth. In the form of the human species the slowly evolving sparkly glowing patterns have reached their maximum state of excitement; a new plane has been reached that is far above what came before. The speed of the movements, the distance of the movements, the pervasiveness of the movements, these set the human species apart as the pinnacle of the evolutionary progression of planetary life. The human sparkly glowing pattern is the pattern of technology; it is a sign that life on Earth has become technological. The whizzing around of humans on the motorways, in airplanes, in underground rush hour, bungee jumping, hand-gliding, the movement of humans to space, to the moon, to the International Space Station; these are all the movements of a technological species.

If one was to view the sparkly glowing pattern caused by the human species *on another planet,* then one could be certain of one thing: *the planet has become technological.* Of this there can be no doubt. So, what is the human species? The human species is that part of planetary life which has become technological. The human species is the maximum state of planetary excitement, as represented by the uniquely human sparkly glowing pattern. The sparkly glowing pattern that is the human species is superior to all of the other sparkly glowing patterns on the Earth because the survival of life on Earth (the survival of *all* of the sparkly glowing patterns) ultimately requires technology. This is why the technological pattern entails such excitement.

You might find it hard to engage with the way that I have presented the above – sparkly glowing patterns! Life on Earth as a single entity! However, such a way of approaching the topic of what the human species is isn't required. Just take time to look at the world around you, just look at it. Look at the humans, look at their activities; look at all of the non-human life-forms that you can find. One thing will surely strike you as so obvious that it barely needs mentioning:

- *The human species is the technological animal*

Despite this being so obvious it rarely gets fully acknowledged and consciously appreciated. This means that the question of whether or not it is a good thing that the Earth has brought forth a technological animal doesn't get the consideration it warrants. One could just brush this aside: *Oh yes, of course, the human species is the technological animal; I have never seen a cat use a mobile phone!* But there is a much deeper truth to be acknowledged. To say that the human species *is* the technological animal is to say that it is the bringer forth of technology *and* that it is this fact which separates the human species from all of the non-human life-forms on the planet. The human/non-human division is forged by the bringing forth of technology.

What makes a human superior to a cat? When we ask this question the answer is very plausibly 'nothing'. The answer might depend on the particular individual human one is using in the comparison. However, when we move to the species level (the sparkly glowing movement patterns of humans and cats) then the answer is obvious. The human species is superior to all species of cat because the human species is the bringer forth of technology.

I don't know if this is clear to you. The point is an important one, so I'll elucidate further. Let us compare a cat to Einstein (I am just using Einstein as an example of someone who we can probably safely assume fulfilled their potential as a human being). I would want to say that Einstein is superior to a cat. Let us compare a cat to a human who doesn't fulfil their potential in any way whatsoever. In this case it is not clear to me that the human is superior to the cat; we just have two life-forms; if the cat is fulfilling its potential more than the human then there are good grounds for saying that the cat is superior to the human. However, our prime focus is not on individuals, it is on the human species. It is obvious to me that the human species is superior to all other life-forms on the planet, and this is because the human species is that part of life on Earth which has become technological.

If you are familiar with long-standing debates concerning what makes the human species superior to the rest of the life-forms on the planet then you will be aware how radical what I have just said is. For, this debate typically does not occur on the species level, but on the individual level. In other words, people literally compare one individual human with one individual non-human planetary life-form and try and come up with an attribute which makes the human superior. The list of things people have come up with is long, and sometimes baffling. People have claimed that a 'superior-making' attribute that a human has, but that the non-human lacks, could be: a soul, awareness, thought, intelligence, language, consciousness, tool use, self-awareness, intentionality, the ability to feel pain, mentality, knowledge of the future, and so on. I think we should reject this way of thinking, this attempt to elevate humans by saying that a human has attribute X but a non-human doesn't have X. We should move to the species level.

I have claimed that the human species is the technological animal. The bringing forth of technology is what elevates the human species above non-human planetary life. Does being a technological species entail any particular attributes which are uniquely human? All species have unique attributes, but if humans have unique attributes which are required for the development of technology, then there are grounds for believing that, in this context, these attributes themselves are superior to the unique attributes of non-human life-forms. It seems to me that being part of a technological species entails a particular 'mind-set': seeing one's surroundings as fundamentally 'other' to oneself, believing that one is superior to one's surroundings, and an ability to rationalise to a high level; it also entails the ability to physically manipulate one's surroundings in intricate ways. A self-propelling and self-reinforcing process can be set in motion because as technology spreads and advances its effect is to isolate and surround, almost to suffocate, the humans that are engulfed by its presence. This bolsters the sense of 'otherness' and of 'superiority'. In short, I think we can say that the bringing forth of technology creates a sense of opposition, and that the human species is therefore *that part of the Earth which considers itself to be not natural*. Indeed, the human invention of the concept of the natural, as something which is opposed to human activities, suggests that this is so.

This sense of opposition seems to be at the heart of humanity. The human species is that part of planetary life which has become technological and the bringing forth of technology creates a sense of 'otherness' concerning the non-human; furthermore, technological development gives rise to an increasing sense of 'otherness'. This sense of otherness/separation is required for the scientific experimentation and intricate technological manipulation of non-human planetary life, and the non-human in general. One could not bear to violently experiment on, to manipulate, to exploit, oneself; so, one conceptualises the non-human planetary life of which one is a part as the 'other'. This estrangement from the planet is a cause of human suffering, whilst at the same time it is a cause of celebration for life on Earth. Our estrangement as a species is necessary but it also prevents us from seeing why there is ultimately a cause for celebration. This is why, as I concluded towards the end of *Chapter Four:*

> *We should be celebrating, yet everywhere there seems to be panic and despair.*

The human species is the bringer forth of wondrous technology and it suffers because it considers itself to be not natural. Part of this suffering is the despair that is caused by the human modifications to the planet and the accompanying belief that these modifications are fundamentally bad. We will consider the relationship between this belief (the 'environmental crisis') and technology in the next chapter. Let us conclude this chapter by briefly considering the relationship between the human species and individual biological organisms.

It is the human species which is the phenomenon of importance to planetary life. Just as we need to think of planetary life as a single entity we also need to think of the human species as a single entity. When we talk of the evolution of human culture we are talking about the way that this single entity unfolds over time. We can think of the human species simultaneously as:

What is the Human Species?

1. The technological animal (that which is engulfed by technology).

2. The sparkly glowing pattern which is the highest state of excitation ever to have existed on the Earth.

3. That part of the Earth which considers itself to be not natural.

So, the human species is not a biological term. It is a term which refers to a particular kind of actions/movements (1), movements/feelings (2) and view of itself (3). The implication of this is that the human species came into existence at the moment that part of planetary life first considered itself to be not natural; this will be the same moment that planetary life became technological; and it will be the same moment that the sparkly glowing patterns reached their highest state of excitation up to that point (as technology unfolds and evolves the state of excitation reaches an even higher level). Before this moment biological organisms existed and, looking back in time, most humans today would be inclined to call these biological organisms 'humans'. However, I am suggesting that we shouldn't think of these organisms as humans; they were fundamentally different from what they became.

In terms of the unfolding of the planet, the existence of just another non-technological organism is of no great significance. It is when (1)(2)(3) comes into existence that an event of immense significance has occurred. I think it is useful to refer to this event as the coming into existence of the human species. (If one prefers to think of the 'human species' as a biological term then one can invent one's own phrase to refer to that which came into existence at the (1)(2)(3) event.)

There is one final implication of this view that I would like to make clear. Let us imagine that two members of the human species that are alive today have a baby. It is possible, in exceptionally unlikely circumstances, that this baby will not become part of the human species. If the baby was, at birth, abandoned in a remote forest, and it lived its entire life in secluded wilderness, isolated from any human contact, then it is likely that this biological organism would not be part of the human species. If it didn't develop its own technology (given that it is isolated from the technology developed by the human species), if it didn't develop a sense of superiority/non-naturalness, if it didn't instantiate technological sparkly glowing patterns, then it wouldn't be part of the human species. Having a certain biological/genetic composition is not sufficient to make an individual organism part of the human species. It is the existence of (1)(2)(3) which delineates the boundary between the human and the non-human.

Chapter 7

Technology and the Environmental Crisis

We have concluded that the human species is the technological animal; the human species is that part of life on Earth that has become technological. The human species is also the initiator of the environmental crisis. In this chapter our aim is to consider the relationship between these two aspects of human existence. Let us start by considering the nature of the environmental crisis. Sloep and Dam-Mieras define an environmental problem as follows:

> any change of state in the physical environment which is brought about by human interference with the physical environment, and has effects which society deems unacceptable in the light of its shared norms.
>
> (Peter B. Sloep and Maris C.E. van Dam-Mieras, 'Science on Environmental Problems', in P. Glasbergen and A. Blowers (eds.) *Environmental Policy in an International Context: Perspectives,* Oxford, Butterworth-Heinmann, 2003, p. 42)

This definition encapsulates a sliding scale of environmental problems from those that are local and temporary on the one hand, to those that are global and long-lasting on the other. The 'environmental crisis' as a concept has arisen because of the emergence, in the last century, of an increasing number of environmental problems that are towards the global and long-lasting end of the scale. The environmental crisis is thus wholly caused by the human actions which have created environmental problems that are characterised by their global reach and long-lasting nature. It is worth highlighting that the environmental crisis is defined by humans as a phenomenon which is wholly caused by human actions. Recall from *Chapter Three* that there are two different types of global warming and that the most fundamental type is *not* resultant from human actions. So, the most fundamental threat to the continued thriving of life on Earth does not arise from the environmental crisis. In order to make progress in our understanding of the environmental crisis let's define some key terms.

- Some Useful Definitions

Sustainability

Sustainability pertains when something exists in a steady state. In the environmental realm the term is generally used to refer either to a resource that humans use or to the state of a part of the biosphere. So, if the resource of fresh water is being replenished in underground aquifers at the same rate as the resource is being used, then it is being used sustainably. When it comes to the sustainability of a part of the biosphere then things can be more complicated due to time-lag effects. So, when it comes to the average global temperature of the atmosphere (the GMST) we need to think of sustainability in two ways. Firstly, we can say that sustainability exists when the GMST is within a certain range (the range required for complex life to survive). Secondly, as the atmosphere is a continuous battleground between opposing forces, we need to consider all of the relevant forces, both human and non-human, which are currently in existence. Some of these forces will be time-lag forces resulting from past events whose effects have yet to become manifest. When we have identified all of the relevant forces we can then calculate whether the effects of these forces will balance out in the future and thereby result in sustainability.

Climate Change

Climate change is a long-term change in weather patterns. Climate change has been a consistently regular and widespread phenomenon throughout the history of the Earth. It occurs today through wholly natural phenomenon such as the El Niño Oscillation. Humans have initiated climate change through activities such as mass deforestation. Whilst humans can be the cause of climate change, they cannot stop climate change from occurring. This is because if one lives on a complexly interconnected planet which is spinning through space in an ageing Solar System, then it is inevitable that the climate will frequently change.

Global Warming

Global warming occurs when there is an increase in the average global temperature of the atmosphere (the GMST). It is normal for the GMST to increase or decrease over time within its 'sustainability range'. What is worrying is the possibility that global warming may occur that pushes the GMST out of its sustainability range; this is worrying because the survival of the humans, and the other complex life-forms which currently exist on the Earth, is dependent on the GMST being within its sustainability range.

The long-term factor which is putting upwards pressure on the Earth's GMST is the increasing output of the Sun (incoming solar radiation to the Earth has increased by 25% since planetary life arose). In the short-term another factor is putting upwards pressure on the Earth's GMST: the human species removing carbon sinks (such as the rainforests) and releasing carbon from its underground storage areas. There are thus two types of global warming – non-human-induced and human-induced.

Environmental Crisis

The environmental crisis is a purely human-induced phenomenon (climate change and global warming both occur in the absence of humans). The human species affects its environment just by existing. However, merely affecting the environment obviously doesn't entail an environmental crisis. The reality is that the human species has affected its environment in a number of very large ways – ways that affect the entire planet and which are very long-lasting. It is this family of very large and very worrying changes which humans have made that as a whole constitute the environmental crisis. This family of human-initiated changes is typically taken to include the following: the mass extinction of other life-forms, climate change, global warming and the weakening of the ozone layer. The environmental crisis is a term which refers both to changes which have already occurred and to fears about likely changes which look set to occur in the future. The environmental crisis is obviously very concerning.

Care for the Environment/the Earth

The human species cares for the environment/the Earth if it acts in a sustainable manner so that the ecosystems and the life-forms of the Earth exist way into the future. The idea that the human species should take care of the environment/the Earth is found in the theological notion that the human species has dominion over the Earth. To care for the environment/the Earth is to have the interests of planetary life as one's objective and to act so as to maximise the wellbeing of planetary life. In concrete terms, what it actually means for the human species to care for the Earth/planetary life is not obvious (recall from *Chapter Two* that there are two paths facing humanity; these paths can be thought of as a 'caring path' and an 'uncaring path').

At one extreme are those who believe that for humans to act in the interests of the Earth and planetary life is for humans to minimise their involvement with the Earth, through reigning in technology, using less resources, emitting less carbon and reducing their population size. At the other extreme are those who believe that for humans to act in the interests of the Earth and planetary life is to increase human involvement with the Earth through technologically controlling the temperature of the atmosphere, this being a requirement for the continued existence and flourishing of planetary life. In this view the human species is the saviour of planetary life and its dominion over the living and non-living resources of the Earth is in the interests of planetary life.

The force to environmental destruction & the force to environmental sustainability

There are two forces at work in global human culture. The overwhelmingly dominant force is the force that has been slowly, but inevitably, increasing in strength for thousands of years; it is a paradigm example of the 'snowball effect'; this force is *the force to environmental destruction*. This force is powered by an escalating global human population and the amount of resources that are used by the individuals in this population. This force is currently so powerful for a number of reasons. Firstly, humans in large parts of the world have a very high and deeply unsustainable level of resource use. Secondly, the human population size is still escalating. These two factors – an escalating population, with a significant proportion of that population with a high resource use per head – are a potent force for environmental destruction. This force is reinforced by a third factor. The majority of people who have experienced living with a high resource use per head have ingrained habits/ways of living, and are typically very reluctant to any suggestion that they should use far less resources; it would almost seem reasonable to suggest that the majority are incapable of making anything other than piecemeal changes to the way that they live. Fourthly, this whole situation is made woefully worse by the fact that both the new members who are born into the escalating population, and the existing members of the population who are living on a very low resource use per head (as in large parts of Africa), have a large proportion of members who, quite understandably, aspire to reach the living standards of those with a high resource use per head. This combination of forces is an almost unstoppable juggernaut heading for environmental destruction.

The second force at work in global human culture is *the force to environmental sustainability*. This force is in its infancy; it is a force which has only had any real power since the second half of the twentieth century. This force is at work when some people feel slightly guilty about their unsustainable lifestyles. It is in operation when some people take an airplane flight to the other side of the world to take an 'eco-holiday' rather than a 'normal' holiday. This force includes much of the recycling that goes on, the reuse of materials, and the development of wind farms and other renewable energy sources. This force is also in operation when people campaign against deforestation and human-induced global warming, and when political leaders meet up to discuss these issues.

Both of these forces are inevitable components of global human culture at this moment in time. At the moment *the force to environmental destruction* is overwhelmingly dominant and this is the way that things need to be. In *Chapter Ten* we will explore the forces which propel the unfolding evolutionary processes on the planet and in so doing we will gain a greater insight into the nature of these two forces. Whilst it is useful to think of these forces as different forces, in reality they are both different elements/expressions of a single force. Furthermore, whilst I have introduced these two forces in terms of how they manifest themselves in the realm of human culture, their scope is much wider this; these forces transcend the human/non-human boundary. Indeed, the evolutionary trajectories of the

Universe/Solar System are determined by this singular force. For the vast part of the history of the Solar System this force has been equivalent to *the force to environmental destruction.* However, at a particular stage in the evolution of the Earth-Solar System this force split itself into 'two' and brought forth *the force to environmental sustainability;* this split happened very recently.

Tool Use versus Technology/Advanced Tool Use

It is very important to appreciate the difference between tool use and technology. Tool use precedes technology and is an ability that numerous non-human planetary life-forms have. The bringing forth of technology on the Earth is a uniquely human ability. Tool use is a central part of being technological, but there is much more to being technological than just tool use. A technological species is able to design tools of immense complexity – tools such as telescopes, mobile phones, satellites, space stations and televisions. The design of tools of such complexity requires in-depth knowledge of the universe. So, a technological species is a species which uses in-depth knowledge of the universe to create complex tools. We can say that technology entails *advanced tool use,* rather than just *tool use.* In one sense, there is clearly a sliding scale which moves from *tool use* to *advanced tool use*; but it is best to think of the difference as a difference in kind, for when advanced tools/technology evolve something very new and very significant has happened which is of a very different kind from that which came before.

It is important to be clear on where the dividing line is between tool use and advanced tool use/technology. The realm of advanced tool use/technology is a broad one, and advanced tools such as space stations, nuclear power stations and iPods are at the currently most advanced edge of that realm. I wouldn't want you to think that much simpler human constructions are not also examples of advanced tool use/technology. I would suggest that the tools used by hunter-gatherers – bow and arrow, harpoon, atlatl, projectile points – are actually advanced tools/technology. These 'advanced tools' are to be contrasted with 'non-advanced' tools such as twigs and stones. I have claimed that *a technological species is a species which uses in-depth knowledge of the universe to create complex/advanced tools.* You might think that the creation of a harpoon and a bow and arrow doesn't require in-depth knowledge of the universe. It is certainly true that the creation of a space station and an iPod requires much greater in-depth knowledge of the universe than the creation of a bow and arrow. However, I think that when a part of planetary life gains sufficient knowledge of the universe to create a bow and arrow something really significant has happened. The technological species has been born; that which sees itself as non-natural, as opposed to the rest of the universe, as superior to non-human planetary life, has emerged on the Earth. So, we should think of the knowledge required to create a bow and arrow as 'in depth' knowledge of the universe, whilst accepting that there are immensely greater depths of knowledge to be attained.

Returning to the theme of the previous chapter, we can say that the human species came into existence when hunter-gatherers brought forth the first advanced tools into existence. Prior to this event there were no advanced tools and there were no humans. Prior to this event there were just tools and life-forms.

- The technology/environmental crisis relationship

Having defined and explored some key terms we can now consider the relationship between technology and the environmental crisis. We can think of these two phenomena as things which emerged at a particular point in the evolution of life of Earth. There is an intimate relationship between these twin emergences. There was a point in the Earth's history when it became technological; there was a point in the Earth's history when the environmental crisis began. The timing of these events is linked. Indeed, the former is the cause of the latter; the emergence of technology results in a chain of events which lead to the environmental crisis. Furthermore, human awareness lags behind this chain of events. In other words, there is a four-stage process. Firstly, humans brought forth technology. Secondly, technology inevitably took on its own momentum and pervaded the planet in a multitude of applications. Thirdly, technology became so pervasive that it caused changes to the Earth that humans would consider to be an environmental crisis (if they were aware of the changes). Fourthly, the human realization dawns that the widespread deployment of technology has resulted in worrying changes that constitute an environmental crisis (for example, humans realize that their advanced drilling technologies have released colossal amounts of carbon from underground storage and that this will raise the atmospheric temperature).

It seems likely to me that this four-stage process will occur on any successful planet where life is thriving. Recall that the 'human species' is that part of planetary life which has become technological. This means that any planet which harbours life which advances to the point of bringing forth technology will contain the human species (the term does not refer to a particular bodily structure or genetic makeup). I say this 'seems likely' because I am open to the possibility that there is a universal memory which enables distantly located parts of the universe to learn from each other. If this is so, then it is at least conceivable that on another planet where life is evolving that stage three and four might be appropriated into stage two; technology would pervade in the full human knowledge of its effects and these effects could therefore be lessened.

Let us recap. I have described the four-stage process which has occurred on the Earth. I have also claimed that the human species is that part of planetary life which has become technological, and that such a becoming is a wonderful event for planetary life. Given that becoming technological is a wonderful event, and that becoming technological leads to an environmental crisis (the four-stage process), we are left with the conclusion that the environmental crisis needs to be seen in a positive light. If we were to imaginatively isolate the changes that constitute the environmental crisis then they seem to be wholly negative. However, we need to see the bigger picture; we live in a complex interconnected world; imaginatively isolating things in this way is not very sensible. In reality, the changes that humans have labelled the 'environmental crisis' need to be seen as deleterious side-effects of a much greater good – planetary life becoming technological.

Technology comes as a 'package deal'; when the technological genie is released from its bottle it inevitably proliferates into all domains of existence on a planet. This is the reality of technology and it is the reality of the history of the Earth. One can look back and think *it would have been better if technology had not leaked into this particular domain.* However, when one seriously considers the way that technology proliferates and spreads once it has been unleashed, then one cannot reasonably expect that in reality this particular technology would not have been created. It is only after the unleashing of technology that one can look back and see the damage that has been done to the environment and the risks that have been created. It is generally only when a technology perturbs a system that one may then come to realise that the technology perturbs the system. The scientific realisation of environmental damage follows the technological perturbation.

The nature of technology as a 'package deal' means that one cannot reasonably pick and choose which elements of technology one prefers. When one looks back at the past one can create a conceptual dichotomy by thinking *medical technology is a gift, nuclear power is an evil,* but the reality is that technology is a package. The question of importance is whether technology *as a whole* is a gift or an evil. Is planetary life better off once the technological genie has been released, or is it worse off?

There are those who believe that things could have turned out differently. Some people believe that humans could have brought forth simple technologies such as the bow and arrow, small boats, and the wheel, and then technological development could have stopped before it reached the stage where serious environmental damage was caused (there is technology but no environmental crisis). This view seems to me to be completely wrong.

Let us accept that technology is a package deal and that when the technological genie is released it is inevitable that technology will continuously pervade and develop. There are different views concerning whether this is a good or a bad thing. There are those who believe that the human species, and life on Earth, would be far better off if the technological genie had remained in its bottle. These 'neo-luddites' now typically state their objection to technology to be the environmental damage that it causes. In the pre-technological era there were impulses to mould and transform the surroundings, but there was no easy means to carry out such mouldings on a large scale. The unleashing of the technological genie enabled such transformations to extend the breadth and depth of the planet. The basis of this environmental objection to technology has two variants. Firstly, humans are unable to control themselves and this means that once armed with technology they will transform the planet to such an extent that it is not a conducive environment for humans to live (humans will bring about their own extinction). Secondly, humans are unable to control themselves and this means that once armed with technology they will transform the planet to such an extent that a great number of the non-human life-forms that live on the Earth will be driven to extinction. This second objection to technology already has a basis in reality – human technologies are causing the extinction of non-human life-forms. It would surely be hard to find anyone who thinks that this side-effect of technology is in itself a good thing.

There are clearly dangers and negative effects arising from the unleashing of the technological genie, both to humans and to non-human life-forms. However, we have seen that technology is a 'package deal' and this means that the negative effects need to be balanced against the positive effects in order to see whether technology is an evil or a gift. I have claimed that the environmental crisis needs to be seen as deleterious side-effect of a much greater good – planetary life becoming technological. In the next chapter we will consider in more depth why the emergence of technology is such a wonderful event for planetary life.

- Two Categories of Environmental Problems

I have described technology as a 'package deal'. When life on a planet becomes technological it gains immense benefits but it also has to endure a painful period of technological birthing in which technology brings a great number of harms. These harms are the 'price to pay' for the much greater immense benefits which technology brings. We can think of the technological birthing process as having three stages.

In the first stage, technology brings both widespread benefits and harms which affect both humans and non-human life. These are very varied. The benefits include medical technologies which can save the lives of human and non-human life-forms, and telecommunications technologies which enable humans to communicate more efficiently. The harms include illness caused by technological pollution and the deaths of multitudes of human and non-human life-forms due to the operations of, and accidents involved in, the use of cars, buses, airplanes, wind turbines and a plethora of other technologies. In this stage it is not clear whether overall technology is 'good' or 'bad'. In the second stage, the technological birthing process reaches its fruition and the immense benefits of technology for planetary life manifest themselves. It is now clear that overall technology has immense benefits. In the third stage, the harms caused by technology are vastly reduced as humans learn to rein in the dangers of various technologies through greater understanding, better design and wiser deployment. These are the three stages of the unfolding of technology.

The environmental problems that constitute the environmental crisis can also be thought of as a 'package deal'. This is because when life becomes technological a vast array of environmental problems will arise, and these problems have the same (technological) cause. However, there is a widespread tendency to mentally package all environmental issues into a single category and to assume that they all have the same kind of solution. At one extreme there are a great many people who believe that, when it comes to environmental issues, technology and human intervention are both wholly bad; for these people the solution to *all* environmental problems involves reducing technological deployment/human intervention. You will hear these people say things like: *How could the solution to a problem possibly be more of what caused the problem?!* At the other extreme are those who believe that technology provides the solution to *all* environmental problems; for these people the solution to all environmental problems is more of what caused the problem in the first place.

I am convinced that these two extreme views, which are surprisingly common and widespread, are both wrong. We shouldn't put all environmental problems into a single category and then adopt a blanket view concerning whether or not technology is the solution. The first thing that we need to do is to create two categories of environmental problems. The first category contains human-induced global warming. The second category contains all of the other environmental problems. By the time that you have reached the end of this book my aim is to have convinced you that this division is an appropriate one, and it is appropriate because the *only* solution to the first category problem is a technological solution. In contrast, all of the second category environmental problems could have either a technological or a non-technological solution. In this category there is scope for choice; we can ponder: *Shall we aim for a technological solution, a non-technological solution, or a mixture of the two, for this particular environmental problem?* We have no such choice when it comes to the realm of global warming – the environmental problem that is human-induced global warming in tandem with non-human induced global warming. There is ultimately, at the end of the day, only a technological solution.

Let us now consider in more depth the range of ways in which life benefits from becoming technological.

Chapter 8

Why Life Benefits From Technology

Technology is very widely seen as something which humans bring forth for the benefit of humans at the expense of non-human planetary life. When we zoom out our focus and consider the total effects of technology on the Earth since it became technological, then this seems to be largely true. Technology has caused an immense transformation of the global biosphere: rainforests have fallen to the chainsaw and mechanised agriculture; concrete jungles have replaced the vegetative habitats of numerous species; human movements around the world facilitated by airplanes and ships have spread 'alien' species and diseases which have caused the decimation of 'native' species; oil spills have decimated oceanic life in particular regions; the oceans have been plundered of their life-forms by human fishing technologies; the technological release of underground stored carbon is affecting the global atmosphere; technology has also led to immense suffering through vivisection and factory farming. Technology has directly caused these changes; it has also indirectly contributed to the escalation of the human population through medical technologies, medicines, and increased food output. Furthermore, technology has also resulted in an increase in the *per capita* consumption of the Earth's resources. Biodiversity has been affected by technology to such an extent that there is widespread talk of a human-induced technologically-driven mass extinction of life on the planet.

All of the above is obviously true. Yet, my aim in this chapter is to convince you that planetary life actually benefits from technology because the benefits of technology massively outweigh the harms. Recall that in *Chapter Three* we considered the phenomenon of non-human-induced global warming and I claimed that the recognition of this phenomenon provides support for the view that Path 2 is the appropriate path that we should tread; more than this, it suggests that actually we need to tread a much more profound path. The future existence on the Earth of the human species and non-human complex life requires:

> **Path 3:** Humans use a whole range of technologies to regulate the temperature of the atmosphere.

In *Chapter Three* we recognised that there comes a point when simply technologically regulating the amount of greenhouse gases in the atmosphere becomes ineffectual; greenhouse gases have become minimised but incoming solar radiation continues to increase. This means that ultimately technologies are required which block the incoming solar radiation from reaching the atmosphere of the Earth in the first place. These technologies are required for the survival of complex life so let's not dilly-dally around; let's push ahead and develop them as soon as we can.

In the previous paragraph I used the phrase "there comes a point". Is it possible to know exactly when in the future this 'point' is? That is: Is it possible to know at which future date the survival of the human species and other complex life is dependent on the deployment of these technologies? There is no crystal clear answer to this question. In attempting to answer it we first need to keep in mind that there are two future requirements for technologies that regulate the temperature of the atmosphere. The most imminent need is for technology to regulate the amount of carbon dioxide in the atmosphere (Path 2). As we extend slightly into the future there is also a need for technology to block the incoming solar radiation from reaching the atmosphere of the Earth (Path 3).

Don't be lulled into the oft-peddled belief that technologies that regulate the atmospheric temperature will only be required millions of years in the future. There are many people who acknowledge the need for technological regulation of the atmospheric temperature if humanity and complex life are to exist on the Earth in the distant future; however, these people assume that this fact (due to non-human-induced global warming) is far removed from the current environmental crisis. In other words, these people wrongly assume that human-induced global warming and non-human-induced global warming are two unrelated independent phenomena operating over very different timescales. They also fail to appreciate the delicate interplay of planetary and Solar-Systic forces which have caused the planetary homeostatic regulatory capacity to be currently in a state of great weakness. The result of these erroneous beliefs is that these people will usually be supporters of Path 1.

The reality of the situation is that there is an imminent need for technological regulation of the atmospheric temperature; that is, there is an imminent need for Path 2 and a slightly less imminent, but also imminent, need for Path 3. The reality of the situation is that human-induced global warming is deeply intertwined with non-human-induced global warming. The weakening of the planetary homeostatic regulatory capacity (which has been offsetting the force that is non-human-induced global warming) has been temporarily exacerbated by human-induced global warming, thereby exacerbating the pre-human need for technological regulation of the temperature of the Earth's atmosphere. We are living at a time at which the evolving interplay between these two forces – human-induced global warming and non-human-induced global warming – has reached the stage at which the need for technological regulation of the atmosphere is nigh. We are talking about possibly within the next hundred years, perhaps hundreds of years, perhaps a few thousand years if we are very lucky. There is a lot of uncertainty about the exact timing of the need due to our ignorance of the state of the biogeochemical cycles of the Earth and the way that these will change in the future. Given that it is possible that the need is within the next hundred years, there really is a need to get our act together. The realisation that there is a definite need for the technological regulation of the atmospheric temperature, and that this need is in the immediate/very near future, should help us to push forward to get the job done as soon as possible. We shouldn't use our inability to know with certainty in which decade, or in which century, the need exists, as a reason for not developing and deploying the required technologies as soon as we can.

Let me expand a little on the subject of 'need'. When I say that there is an imminent need for technological regulation of the atmospheric temperature I am using the word 'need' in the extremist of senses; without such regulation the human species and all complex life on the Earth will become extinct. Without such regulation the temperature of the Earth's atmosphere will jump upwards making the planet uninhabitable for complex life-forms. The planet would take a gigantic step backwards to its infancy; only the simplest of life-forms would exist. Furthermore, the new planetary dynamics – a much warmer atmosphere combined with the continuing increase in incoming solar radiation – would mean that it was only a matter of time before all life on Earth became extinct. This is because, following the jump in atmospheric temperature, the homeostatic regulatory capacity of the Earth would be irreparably weakened. Recall that the homeostatic regulatory capacity is already in a severely weak state, this is precisely why technology is required to regulate the atmospheric temperature in the imminent future. If the required technology is not deployed the atmospheric temperature will jump upwards and the ongoing battle between the Sun and life on Earth will forever be lost; the Solar System will have reached the stage in its evolution in which it is doomed to be lifeless (to return to our earlier discussion, the 'Earthly womb' would have had an 'abortion'). Conditions on the Earth would forever more be too hot for complex life; complexification to another technological saviour species would not be possible.

You might not be convinced by what I have just said; you might find it to be alarmist, or too pessimistic. If this is so, then you can consider the imminent need for technological regulation of the atmospheric temperature as follows. The current exacerbation of non-human-induced global warming by human-induced global warming will lead to a plethora of undesirable outcomes if technological regulation of the atmospheric temperature is not implemented in the immediate future. For example, there are changes in biogeochemical cycles 'locked in' for the rest of the century and beyond which will lead to significantly rising sea levels *if* carbon is not technologically removed from the atmosphere (or if other alternative technological means of atmospheric temperature control are not deployed). So, if numerous states and cities are not to become submerged under the oceans, if hundreds of thousands of humans are not to be displaced from their homes, there is a need for technological regulation of the atmospheric temperature. Due to the 'locked in' future changes reducing emissions now will not avert this outcome; technological control is the only solution. The same is true of climate change and the increasing numeracy and extremity of extreme weather events caused by a warming atmosphere. These cannot be averted through emissions reductions, but they can be averted through technological control of the atmospheric temperature. The tragic consequences outlined in this paragraph constitute a less extreme version of the need for technological atmospheric temperature control. Here the need is to avert tragic consequences which affect *certain parts* of the human species and *certain parts* of non-human planetary life.

It should be stressed that the technological control of the atmospheric temperature is not something that you should be overly worried about. We have already used technology to move carbon from underground storage to the oceans and atmosphere. Now we simply need to use technology to move carbon in the reverse direction – from ocean-atmosphere to underground storage. Such a simplistic form of technological regulation, simply reversing what we have already done, is all that is required in the very immediate future. *This reversal leads to stabilisation, not destabilisation. It is the prospect of a lack of technological control of the atmospheric temperature which you should be deeply concerned about.*

The main way in which planetary life benefits from technology is that technology enables planetary life to exist after a certain stage in the evolution of the Earth-Solar System. Planetary life has been positively thriving for millions of years without technology, but we cannot make a simplistic extrapolation from the past to the future. We cannot say because this was so in the past it will also be so in the future. Such a way of thinking can only take hold in a person who simply does not comprehend that the Earth-Solar System is an ageing/evolving entity, a person who does not see that the Earth is currently in a very delicate state of balance in its ability to support complex life due to the massive increase in the Sun's output. It is obvious that the future existence of complex life requires the deployment of technology to keep the temperature of the atmosphere suitable for its existence.

The development of technology causes changes to the composition of planetary life, changes such as the extinction of some life-forms. The development of technology also causes suffering to both humans and non-human planetary life. These are grounds for believing that technology is a bringer of harm to the planet. Yet, without the necessary deployment of technology, planetary life will soon be decimated. First of all the complex life-forms would become extinct as the temperature of the atmosphere jumps upwards (forever). In this stage non-complex life-forms would still manage to survive in the hostile conditions, but with the increasing solar output and the collapse of the planet's homeostatic regulatory capacity, a re-emergence of complex life wouldn't be possible; the Earth-Solar System would simply have evolved into a state/age where this is unable to occur. It would be only a matter of time until the Earth returned to a lifeless state. Life would just have to try again on another suitable planet, and 'hope' that it could survive long enough for technology to both evolve and be more intelligently deployed than it was on the Earth (i.e. hope that technology is deployed to regulate the atmospheric temperature).

So, whilst technology entails harm and suffering, it also becomes the enabler of the very existence of precious life. It is better for there to be existence enabled by previous harm, than for there to be no harm and no existence. Surely, all can agree on this. The technological birthing process on a planet is a process that involves great harm. Yet, it is required; the requirement for technological regulation of the Earth's atmospheric temperature in order for planetary life to survive and thrive is non-negotiable. As we have explored, the only doubt is about the timing of the need; there is no doubt about the need itself.

How else does planetary life benefit from technology? As we extend into the very distant future, technological control of the atmospheric temperature will not be sufficient to save life on Earth; the expiration/explosion of the Sun will mean that the survival of the life which has evolved on the Earth will only be possible if it relocates to a different Solar System. There are two ways in which this could occur and both require very advanced technology. These technologies might seem a bit far-fetched to most people who are presently alive. The most outlandish possibility entails technologically moving the Earth to another Solar System. The more plausible possibility seems to me to be the creation of an advanced spacecraft which could be populated with a great variety of the life-forms which exist on the Earth. This spacecraft could plausibly leave the Solar System before the expiration of the Sun. This variety of life-forms, which survive the end of the Solar System, could either repopulate another planet or make their spacecraft their new home (it is likely to be super-luxurious compared to our present infantile spacecraft). Without the utilisation of technology in one of these two ways the distant future entails death for the life that has arisen on the Earth.

There are also other immediate ways in which planetary life benefits from becoming technological. One of the most obvious ways is that technology can protect planetary life from potential annihilation by a massive asteroid collision with the Earth. A smallish asteroid strike wouldn't threaten planetary life, it would just cause suffering to some parts of planetary life; it could also possibly change the composition of planetary life through causing the extinction of some species. However, a sufficiently large asteroid strike could threaten planetary life in its totality. This possibility is enhanced when one realises that the ability of planetary life to successfully homeostatically regulate the conditions required for life is already under stress. A major perturbation such as a large asteroid strike could be a catalyst for the collapse of an already stressed system, and it could result in the end of complex life. Thankfully, technological development programmes are already well under way which have the objective of deflecting or destroying asteroids which are on a collision course with the Earth. Two of these programmes are NASA's Near Earth Object Program and the European Union funded NEOShield project. There are a vast array of technological collision avoidance strategies being considered – nuclear explosive devices, kinetic impactors, asteroid gravitational tractors, and ion beam shepherds, to name but a few. Through bringing forth such technology planetary life is bolstering its defences and doing everything that it can to help ensure its future survival.

Another possible threat to planetary life is the eruption of a massive supervolcano (according to scientists an explosion of the Yellowstone

Supervolcano is overdue). It is at least plausible that such an explosion could pose a threat to the continued existence of planetary life. Technology can help to avert this calamitous outcome. If humans construct a large 'biome' with an artificial atmosphere (a kind of stationary Ark, or even something which circulates in the atmosphere) which is self-propagating and isolated from the atmosphere and workings of the rest of the Earth, then planetary life could survive such an explosion. If the life-forms of the Earth existed within such a technological creation then they could survive the change of planetary conditions brought about by the supervolcano explosion. If the external planetary environment returned to a favourable state then these life-forms could repopulate the planet; if not, then they could continue to live within the biome.

Another possible threat to life on Earth, which is hard to assess for its plausibility, is that life on Earth could be invaded by aliens which seek to destroy life on Earth. These aliens could attempt to use their technology to wipe out life on Earth and then move in and live here themselves. If life on Earth is technologically prepared, through the human species, then it might be able to successfully defend itself from such an invasion. Needless to say, if this situation was to pertain, and the human technology was successful, then the rest of life on Earth would be 'grateful' to its technological saviour.

There are clearly lots of ways in which life on Earth benefits from becoming technological. The benefits of technology for planetary life massively outweigh the harms which are created in the process of becoming technological. I am sure that you can now understand why technology is a wondrous gift for life on Earth.

Chapter 9

Is the Damage Already Done?

In the previous chapter I outlined why planetary life benefits from becoming technological. We are living at a time when planetary life has become technologically advanced, yet technology hasn't yet advanced to the stage where the benefits of technology for planetary life have come to fruition. We are living at a time of technological birthing, and birthing is a painful process; the harms of technology are apparent, but the wonderful fruition of this painful birthing process is yet to come.

Is it possible that the present reality of technological harm could cause the human species to turn its back on technology? Could the human species initiate an abortion thereby scuppering the wonderful birth that is on the imminent horizon? I don't believe that such an abortion is possible. There are two reasons which underpin this belief. Firstly, I believe that the evolutionary processes which have bought planetary life to this point, from the first stirrings of life in the primordial soup to globalised technological society, are very powerful. These forces underpin both the motivations/actions of individual humans and the evolutionary trajectory of human cultural development. These forces will ensure that the human species does not initiate an unwanted abortion. Secondly, I believe that it will soon be widely recognised, on the rational level, that we don't have any choice but to push ahead with the technological regulation of the Earth's atmospheric temperature. We don't have any choice because the biogeochemical cycles of the Earth have been perturbed to such an extent – through the two types of global warming that we have explored – that the only solution to the problematic situation that we face is a technological solution. In short, the damage has already been done and technology is the only solution. In this chapter we will focus on this second reason, the claim that the damage has already been done. In the next chapter we will explore the first reason, the evolutionary processes which propel the planet.

In the *Introduction* we saw that there are two ways in which humans might be able to stop carbon dioxide concentrations in the atmosphere from rising too much:

> **Path 1:** Humans stop emitting, or radically reduce emissions of, carbon into the atmosphere.
>
> **Path 2:** Humans use technology to regulate the amount of carbon dioxide in the atmosphere.

I stated that according to one line of thought Path 1 is not even an option. This line of thought has two strands. Firstly, the damage has already been done, simply stopping now will have no effect; the action-consequence time-lags mean that carbon dioxide concentrations are set to keep on rising for the foreseeable future whatever we do now. Secondly, we simply cannot stop emitting now; the state of the world (population size and growth, economic trajectories, developing countries industrialising, state of technology, individual motivation and desire) and the human dependency on cheap fossil fuel energy supplies means that Path 1 is nothing more than fanciful wishful thinking. In other words, according to this second strand of the line of thought, the damage hasn't fully been done yet, we are not yet past the point of no return, but we inevitably will be very soon. In this chapter I will consider this line of thought in greater depth.

There are three broad groups of opinion concerning the human response to *human-induced* global warming:

Group 1: Humans should reduce their use of resources, recycle more, reuse more, possibly reduce/control their population size, leave things to 'nature' as much as possible, and use renewable 'green' (non-nuclear) energy sources.

Group 2: The yet to be manifested effects of previous perturbations to the biogeochemical cycles caused by humans, and/or current social realities, mean that the only way to alleviate global warming is to dash for nuclear power as soon as possible. If we listen to *Group 1* advocates the result will be catastrophic global warming; this will be to the detriment of life on Earth.

Group 3: Even a dash for nuclear is not enough. The damage that has already been done is far too serious for this. What is required to stop catastrophic global warming is the active technological regulation of the temperature of the atmosphere by the human species. If we listen to *Group 1 or Group 2* advocates the result will be catastrophic global warming; this will be to the detriment of life on Earth.

Group 1 is obviously an expression of Path 1. Group 2 is also an expression of Path 1. Group 3 is an expression of Path 2. It is the Group 3 opinion which I will be outlining and defending in this chapter. In other words, I will be making the case that treading Path 1 is a futile exercise. That is to say, reducing greenhouse gas emissions is a futile exercise; it isn't a plausible remedy to the global warming problem. The reason that this is so is that the damage has already been done. It doesn't matter if we reduce emissions now, or even if we stop emitting altogether, this wouldn't avert the need for technological control of the atmospheric temperature.

As we have explored, long before humans evolved, carbon from the atmosphere was stored under the land surfaces of the Earth in order to keep the temperature of the atmosphere down in the face of increasing solar output. Humans have released an enormous amount of these stores of carbon back into the biogeochemical cycles of the Earth but, so far, the effects of this on the atmospheric temperature have barely manifested themselves. As Stephen Peake explains:

> Even if we engage in further actions to reduce GHG emissions, the models tell us that anthropogenic climate change will continue for centuries.
>
> (Peake, S., 2003. A citizen's guide to climate science. In Peake, S., and Smith, J. eds. 2003. *Climate Change: From Science to Sustainability.* Milton Keynes: The Open University. p. 78)

One of the main reasons for this time-lag effect is that colossal amounts of carbon that humans have released have become temporarily stored in the deep ocean thermohaline circulation, and this carbon will eventually resurface. The thermohaline circulation is a process which involves dense cold water sinking at high latitudes and travelling very slowly through the ocean depths until it eventually reaches the northern Indian Ocean and the northern Pacific Ocean where it resurfaces. As Robert H. Stewart explains:

> New CO_2 is released into the atmosphere when fossil fuels and trees are burned. Roughly half of the CO_2 released into the atmosphere quickly dissolves in the cold waters of the ocean which carry it into the abyss…therefore temporarily reducing atmospheric CO_2. Eventually, however, most of the CO_2 must be released back
>
> (Stewart, R. H., 2009. *Introduction to Physical Oceanography*. Florida: Orange Grove Texts Plus. p. 214, p. 233)

After the carbon has sunk into the thermohaline circulation it takes from 100 years to 1000 years to re-enter the atmosphere. This means that almost all of the carbon that has become stored in the thermohaline since the start of the Industrial Revolution has yet to re-enter the atmosphere. When this carbon starts to be released *en masse* then the temperature of the Earth's atmosphere is likely to shoot upwards. The initial increase in atmospheric temperature could easily trigger other large-scale discontinuities such as a runaway greenhouse effect resulting from the destabilisation of methane clathrate reservoirs. By the year 3000 – when all of the carbon that is currently stored in the thermohaline has been released – the conditions of the Earth could very easily not be suitable for human habitation or for habitation by any complex life. To repeat the point, this applies even if humans were to stop using fossil fuels today; our continuing intoxication with fossil fuels is just marginally bringing forth the 'planetary inhabitability date'. Reducing greenhouse gas emissions, reducing fossil fuel use, will not do anything to avert this scenario. The damage has already been done, and all that reducing current and future emissions will do is possibly buy a little more time. Such actions (Path 1) are not the solution to the problem we face.

The realisation that the damage has already been done, that the solution to the situation we face cannot be emissions reductions, is obviously not widespread. That this is so, or even the possibility that this might be so, barely gets mentioned in the media. There is clearly big trouble in store when all of the carbon that has become temporarily stored in the oceanic thermohaline circulation returns to the atmosphere. We need to be technologically prepared so that we can regulate the atmospheric temperature before this happens; we need to be able to prevent the temperature from rising too much. If we aren't technologically prepared, and we are still alive, then we will really know what 'extreme weather' is!

The media, and the politicians, are totally obsessed with the paradigm of 'carbon emissions reductions'. In reality, the damage has already been done, but this is not a cause for despair. From later this century, until the year 3000, the carbon which has been stored in the thermohaline since the start of the Industrial Revolution will start gushing out of the thermohaline and into the atmosphere. However, this doesn't mean apocalypse for the human species and the rest of complex life on Earth. We know that this danger is on the horizon and we can be technologically prepared to deal with it. There is no other solution; there is no non-technological way to deal with this situation. We need to be prepared; we need to be competent at technologically regulating the temperature of the atmosphere as soon as possible. Some trial and error might be required before we get it right, but we need to get it right before the carbon really starts to gush into the atmosphere from the thermohaline in large quantities. If we are adequately prepared then when the carbon emerges from the thermohaline we can maintain the temperature of the atmosphere for the benefit of life on Earth; indeed, for the very survival of complex life on Earth.

We have so far been considering the future events which will occur due to time-lag effects resulting from human actions over the past few hundred years. Our present actions across the planet are exacerbating the situation and building up even more serious time-lag effects for the second half of this millennium. And we simply cannot stop ourselves from emitting massive amounts of carbon, our fossil fuel and energy intensive lifestyles are deeply ingrained. The state of the world – population size and growth, economic trajectories, developing countries industrialising, state of technology, human motivations and desires – means that the non-technological future looks increasingly bleak. It is to be hoped that those who oppose technological regulation of the Earth's atmospheric temperature come to appreciate the need for such regulation based not only on these 'state of the world' factors, but also on the realisation of the very serious 'time-lag effects', particularly relating to the thermohaline circulation, which Path 1 is impotent to address. Path 1 is nothing more than fanciful and misplaced wishful thinking.

To reiterate what I said earlier, to believe that 'the damage has already been done' can be to believe one of two things. Firstly, it can be to believe that if the human impact on the planet immediately fell to zero (no future human greenhouse gas emissions) that this would make no difference whatsoever to the situation we face. Secondly, it can be to believe that an immense amount of damage has been done but that an immediate massive reduction of global human greenhouse gas emissions could be a feasible solution to the situation we face; however, given the 'state of the world' factors outlined in the previous paragraph, such a massive reduction is simply not going to happen. In other words, the point of no return has not been yet been passed but it is about to be passed, and nothing can stop it from being passed. Both of these subtly different beliefs concerning 'the damage has already been done' lead to the same conclusion: Path 1 is fanciful and misplaced wishful thinking.

In this chapter we have focused on the yet to be manifested time-lag effects caused by human activities. In other words, we have focused on the type of global warming that is human-induced global warming. We need to keep in mind that these time-lag effects are superimposed onto biogeochemical cycles which were already in a state of stress due to non-human-induced global warming. In other words, the 'damage' that has already been done by humans needs to be seen in the context of the non-human 'damage' that has already been done to the homeostatic regulatory capacity of the Earth by the Sun. What was damaged has simply become more damaged.

Technology is the only solution to this damage. Let us prepare our technology; let us heal the damaged Earth.

In this chapter we have been considering the way that the oceans are acting as a global warming buffer. In order to prevent the atmospheric temperature from rising to a level that is not conducive for the survival of complex life, the oceans have pulled down an enormous amount of carbon from the atmosphere into the slow-moving deep ocean thermohaline circulation. This process is part of the homeostatic regulatory capacity of the Earth. The planet is doing everything that it possibly can to maintain the atmospheric conditions which complex life needs in order to survive and thrive. We need to see the movement of carbon to the thermohaline as the Earth's way of giving the human species the maximum time possible to fulfil its purpose of technologically controlling the atmospheric temperature. We need to realise that this is so. The planet is doing everything that it can to help us, to buy us time, but we don't have time for dillydallying around; the non-technological planetary homeostatic regulatory capacity is in a state of great fragility.

In order to avoid confusion, I would like to finish this chapter by mentioning a second short-term time-lag effect which involves the oceans acting as a buffer against a rising atmospheric temperature. This effect has recently received some media attention. In the *i* newspaper in February 2014, Steve Connor outlines this effect in relation to the apparent global warming 'pause' (the slowdown in the increase in the GMST since 1998) which is often referred to by global warming sceptics:

> The easterly trade winds of the Pacific have increased significantly over the past two decades and as a result are blowing higher volumes of warm surface sea water and huge amounts of surface heat down to deeper depths of the ocean, scientists said.
>
> "This hiatus [in the increase of the GMST] could persist for much of the present decade if the trade winds continue," the scientists say in the journal *Nature Climate Change*.
>
> The cooling capacity of the Pacific is not expected to continue much beyond 2020, when global surface temperatures are expected to start rising again rapidly.
>
> *(Cooler Pacific masking global warming effects,*
> Steve Connor, the *i*, 10 February 2014, p. 22)

This is a very short-term effect. It started two decades ago and is expected to be over within the next decade. This is clearly very different to the thermohaline mechanism which we have been considering, which operates over a timescale of hundreds of years. The existence of this short-term effect is clearly potentially very concerning. Without this additional effect things were already looking very bleak because of the time-lag buffering/masking effects resulting from the storage of masses of carbon in the thermohaline circulation, effects which will start to become manifest later this century. Now on top of this we have this additional short-term level of masking. The more we discover about the workings of the biogeochemical cycles of the Earth, the more we are discovering how dire the situation we are currently in is. The forces which exist for future global warming – forces whose effects are yet to be manifested because they are currently masked through time-lag effects – are immense.

If this short-term ocean buffering effect ceases by 2020, then the potential upwards pressure on the atmospheric temperature which will exist in 2020 is clearly a matter of great concern. However, if this pressure results in some deleterious effects, such as a significant increase in the GMST and a wave of extreme weather events, then at least we will have been properly warned. We will then have sufficient time to get technologically prepared for the 'main event', the gushing of the masses of carbon out of the thermohaline into the atmosphere later in the century. It is worth repeating what we concluded earlier in the chapter:

Technology is the only solution to this damage. Let us prepare our technology; let us heal the damaged Earth.

Chapter 10

The Evolutionary Processes Which Propel the Planet

In the previous chapter I outlined one reason why I believe that the human species will not initiate an abortion of the wonderful technological birth that is on the imminent horizon. The technological regulation of the atmospheric temperature will go ahead because the damage has already been done; humans will soon widely come to realise that this is the case and will therefore conclude that the wonderful birth must go ahead.

It is possible that you might not have been convinced by what I said in the previous chapter. You might still believe that the damage has not been done, that Path 1 is a feasible solution to the situation that we face. If you believe this then you can still accept the fundamental philosophical worldview that I am outlining in this book. The road to technological regulation of the atmospheric temperature that I outlined in the previous chapter is based on the conscious awareness of the need for such regulation widely arising in the human species. However, such awareness is not required for a successful birth. Indeed, the fundamental forces which are pushing us towards a successful birth have nothing to do with conscious awareness. It is the evolutionary processes which propel the evolutionary trajectory of life on Earth, processes which operate on the subconscious level, which will ultimately bring about the successful birth. In other words, if you were not convinced by the previous chapter that is fine; if you are convinced by this chapter it will take you to the same destination. The same is true in reverse; if you find the claims I make in this chapter to be totally unconvincing, but found the claims of the previous chapter compelling, then that is fine. Many roads end up at the same destination and it is okay if one is unable to perceive some of these roads. The main thing that I hope is that you find enough to believe in to end up at the right destination.

The evolutionary processes which propel the planet are forces which affect the entire planet. These forces forged the planet in the first place; these forces led to the origin of life on the planet; these forces are responsible for the evolution of life from the one-celled life of the primordial soup to highly complex multicellular life-forms; these forces propelled human cultural evolution from hunter-gather to globalised technological society. These forces are clearly very powerful! These forces underpin both the motivations/actions of individual humans and the entire evolutionary trajectory of human cultural development (in *Chapter Seven* I conceptually divided these forces, as they manifest themselves in the human cultural realm, into *the force to environmental destruction* and *the force to environmental sustainability*). These forces will ensure that the human species does not initiate an unwanted abortion; we cannot disobey our own motivations.

These forces have no concept of 'inner' and 'outer'; they pervade everything. In other words, these forces are instantiated inside human bodies and form the ground which stimulates human actions; they also exist throughout planetary life, throughout the Earth, and throughout the Solar System. One is obviously most familiar with the direct experience that one has of the forces/motivations that propel one from within. These forces/motivations can be described simply as 'feelings' which pervade the human body. A human can be thought of as being constituted by a feeling part and a thinking part; it is the feeling part which we are here primarily concerned with. The feeling part dominates human actions because if the thinking part of a human regularly disobeys the feeling part then that human will be unhappy/miserable. The vast majority of humans prefer to be happy/content/feel good, rather than to be unhappy/miserable/feel bad, so the vast majority of humans typically live and act in accordance with their inner motivations/feelings. When this translates to human cultural evolution, the summation of individual human actions gives rise to a cultural trajectory which is determined by the motivations/feelings that exist within individual humans.

In other words, the vast majority of humans in the past, and today, generally act in a way that seeks to make their state of feeling good rather than bad. Humans actively seek to spend their time acting in ways which make them feel good and not bad. Some of these acts will span the human species. For example, the vast majority of humans will not willingly and consciously walk through a bonfire or get run over by a car. They won't do this because they realise that such actions will result in them feeling very bad (excruciating pain) rather than feeling good.

There are another group of acts which are more specific, these acts will make a particular human feel good, but they will not have the same effect on most other humans. For example, as a particular human matures they might come to realise that doing charity work makes them feel good; it is an activity which satisfies their inner motivational state. One would expect this particular human to do lots of charity work. Another human might feel exhilarated when engaged in complex mathematics, or painting, or architecture, or singing, or designing nuclear power plants, but feel depressed when they are doing charity work. You wouldn't be surprised to find this particular human engaged in one of their feel-good activities, but you would be surprised to find them doing charity work every Saturday morning. The 17th century philosopher Baruch Spinoza clearly saw that this was so. He asserted that humans simply cannot deprive themselves of those things which they judge to be the most conducive to their own welfare (those things which 'optimise' their feeling states). According to Spinoza this is a principle which is:

> inscribed so firmly in the human breast [that it constitutes an] immutable [truth that] nobody [can] ignore.
>
> (Baruch Spinoza, *Tractatus Theologico-Politicus* [1670], Emilia Giancotti Boscherini (Ed.), (Torino, 1972), I, p. 472)

Individual humans acting in a way that is in accordance with what they consider to be most conducive to their welfare (acting in a way that 'optimises' their feeling states) results in the cultural trajectory from hunter-gatherer to globalised technological society. It is important to realise that these feeling states don't just exist in humans. The ancestral species of the human species (the life-form which the human species evolved out of) was not wholly devoid of feeling states! Furthermore, dogs, cats, elephants, chimpanzees, bonobos, dolphins, birds, these parts of planetary life are not wholly devoid of feeling states! Feeling states pervade planetary life and just as feeling states are the forces which propel human cultural evolution along its trajectory, feeling states are the force which propels the entire course of planetary life from primordial soup to globalised technological society. It is one long process of states seeking to act/move in accordance with what maximises/optimises their feeling. Move away from bad feeling and towards what feels good; stay there, stay in the good as much as possible. This is such a simple process, a process which propels planetary life onwards and upwards.

Do these feeling states just pervade planetary life or do they pervade the planet and the Solar System? In other words, did they simply spring into existence in the primordial soup out of arrangements of stuff that was wholly devoid of feeling states? Whether we think about this question on an intellectual level or on an intuitive level, the conclusion that feeling states pervade the universe is a conclusion that I believe we should embrace. On the intellectual level the coherence of the postulated *emergence* of feeling out of that which is wholly non-feeling can be doubted (for example, see the work of Professor Galen Strawson, and his book: *Consciousness and its place in nature: why physicalism entails panpsychism*, Imprint Academic, 2006). On the intuitive level it really seems right that this is so. Feeling states can be seen as the ever-present driving force of all that is. Feeling states underpin chemistry, physics and biology; they thus have a role to play in everything from the formation of the Earth to the bringing forth of the human species. Universe-pervading feeling states can provide an explanation of why things move/act in the way that they do; science simply measures the exterior of things, it simply measures movements/actions. We live in a feeling universe.

I am trying to get you to see that the entire universe is a cacophony of states of feeling which are continuously in a process of seeking to optimise their state of feeling. Furthermore, I am suggesting that this process is the evolutionary process which propels the evolution of human culture, planetary life, the Solar System and the Universe (this evolutionary process is the singular force which I have conceptually divided into *the force to environmental destruction* and *the force to environmental sustainability*). This raises the question of whether there is an end-point to this process. Does the evolution of the universe have a goal, an objective? To say that the universe has a particular objective/goal/purpose (I will use these words interchangeably and to mean the same thing) is simply to agree with Aristotle that:

> There is something divine, good, and desirable... [that matter] desire[s] and yearn[s] for
>
> (cited in Skrbina, 2005, *Panpsychism in the West*. London: The MIT Press. p. 46)

So, the objective/purpose of the universe could be to attain a "good and desirable state", to attain that which is "yearned for". You will notice the similarity between Aristotle's claim and the view of the feeling universe that I have outlined:

> *A 'good and desirable state' is a state of optimal feeling, and the 'yearning for' is the driving force, the motivation, which propels states of the universe to attain more optimal states of feeling.*

If the universe has a purpose, then its purpose has to be to reach a particular state; let us refer to this state as U*. Given that we live in a feeling universe, U* is seemingly, by definition, the state of the universe in which the universe is in a maximal state of feeling. So:

U* = states of optimal feeling exist throughout the universe

What are these states of optimal feeling? What would a U* universe look like? We need to think of the feeling states that pervade the feeling universe as existing within a hierarchy. The states at the top of the hierarchy are states of optimal feeling – intense, exhilarating, exceptionally feel-good states; the states at the bottom of the hierarchy are states of base/minimal/inferior feeling. There is a gradation of states within these extremes. The base states of minimal feeling are the non-living states of the universe. If these states are successful in their yearning for greater, more optimal, feeling they will turn into simple/non-complex life-forms, which are the next level of feeling up the hierarchy. The states of feeling that are simple life-forms yearn for the greater, more optimal, states of feeling that come from being complex life-forms; if they are lucky they will transform themselves in this way and move up the hierarchy further to the complex life-form level of feeling. States of complex life yearn for an ever greater intensity of feeling, for feeling that is more optimal, feeling that is good, more vibrant, more satisfying; in other words, they yearn for that form of complex life that is the technological animal, the human species. The human is the optimal state of feeling which sits at the top of the hierarchy. This obviously means that:

U^* = humans exist throughout the universe

This might seem a bit ridiculous! The purpose of the universe is for a human to exist at every single point in the universe! I don't think this is as ridiculous an idea as it might first appear. To say that the purpose of the universe is to reach U^* is not to say that the universe will actually attain U^*. It is common for purposes to not be achieved – for objectives to not be attained. To say that the purpose of the universe is U^* (humans exist throughout the universe) is simply to say that the entire universe is yearning for, pressing forwards toward, life, complex life, and the technological form of complex life. It is this yearning which characterises the nature of the universe, rather than the actual state of the universe (I don't believe for a second that the entire universe will ever be constituted solely by humans).

This way of thinking leads to the conclusion that the more states of good/optimal feeling there are in the universe, the better state the universe is in. In other words, we should think of U^* as a theoretical goal and the closer to this goal the actual universe is, the better the state the universe is in. So, if a single life-form is born the universe moves closer to U*. If a single life-form dies the universe moves further away from U*. If a planet in the universe brings forth life this moves the universe closer to U*. What joy! If this planet evolves complex life the universe moves closer to U*. How wonderful! If complex life on this planet evolves into the technological form then the universe will move even closer to U*. How glorious such an evolution is!

The technological form of complex life is the optimal state of universal feeling for a very good reason. All life-bearing planets will seek to attain this state of feeling, this technological form, through their yearning for more optimal states of feeling. When this form is attained and has matured then it is the enabler of universal joy, a catalyst for the existence, pervasion and multiplication of states of greater feeling. The technological form can radiate life out into the universe. In other words, the technological form brings forth technology which enables life to exist in space stations, on other planets and ultimately in other solar systems. Closer to home, as we have already been exploring, the technological form is required so that complex life can continue to exist on the planet where it evolved (due to the threat to its existence posed by non-human induced global warming). We explored the range of benefits that accrue to planetary life when it attains the stage of technological birthing in *Chapter Eight*.

I am suggesting that built in to the very make-up of the Solar System and the Universe are mechanisms which give rise to evolutionary trajectories which lead towards U*; these mechanisms are grounded in the feeling and yearning processes that I have outlined. In other words, things evolve according to a roughly predictable trajectory. A particular movement along this trajectory towards U* will be the fulfilment of what we can call a 'mini purpose'; the fulfilment of a 'mini purpose' will simultaneously be the attainment of a more optimal state of feeling for the universe and a stepping stone in the overall evolutionary trajectory of the Solar System/Universe.

You will probably be aware that contemporary conventional wisdom denies that this is so. It is supposedly intellectually responsible to believe that the human species is just one species amongst many, and concordantly that there is no pinnacle or zenith to the evolutionary progression of life on Earth; the human species simply evolved as a fluke. As Professor Stephen Hawking puts it:

> We are just an advanced breed of monkeys on a minor planet of a very average star.
>
> (*Der Spiegel,* 17 October 1988)

Those who peddle this *'we don't have a special place in the cosmos, we are just advanced monkeys'* view, curiously, seem to believe that Darwin's theory of natural selection is an adequate explanation of the evolutionary trajectory of planetary life; as if Darwin had the final word on the matter, rather than one of the first. How limited their understanding is! Imagine that the evolution of life on Earth is a tree which grows from a miniscule size to a massive height. Natural selection gives an explanation of the branches and the leaves, but it gives no explanation of the heart of the tree – the trunk. As the trunk of a tree moves upwards towards the sky it provides a direction from which the branches can spread out. As with the tree, so it is with the evolutionary progression of life on Earth. The evolutionary trajectory of life on Earth is determined by nothing other than the Solar System itself. The fundamental (feeling) nature of the universe (of which the Earth and life on Earth are parts) in tandem with the swirling of the planets around the Sun (as they enter into particular relations with the Earth), determines the evolutionary trajectory of life on Earth. This is the trunk; natural selection is just the branches. How could anyone seriously believe that the entire evolutionary trajectory of life on Earth is determined wholly by natural selection?! This is like focusing solely on the icing in blissful ignorance of the cake.

The trunk of the evolution of life on Earth extends backwards to the time before life on Earth existed. In other words, the feeling mechanisms which fundamentally direct the evolutionary trajectory of planetary life (including human culture) are the same mechanisms which brought life into existence, the same mechanisms which forged the Earth itself. Let us ignore the icing that is natural selection and consider the mechanisms that are the cake.

There are two evolutionary mechanisms which constitute the trunk of the evolutionary trajectory of both life on Earth and non-life on Earth. The first mechanism is the feeling/yearning nature of the universe. The second mechanism is the structure, and interconnected nature, of the Solar System. Let us further consider the first of these mechanisms. Earlier in this chapter I outlined a view according to which the universe is pervaded by states of feeling which are yearning to attain a more optimal state of feeling. This simple mechanism leads to the coming together of atoms, the formation of cells, and the formation of animals; overall it gives rise to a particular evolutionary trajectory which over time leads towards the optimal state of feeling that is the technological form of life. In the realm of the evolution of planetary life this feeling mechanism is concordant with the view of evolution proposed by Professor Lynn Margulis and Dorion Sagan, who have proposed a symbiogenetic theory of evolution. They claim that:

> As members of two species respond over time to each other's presence, exploitative relationships may eventually become convivial to the point where neither organism exists without the other. Long-term stable symbioses that leads to evolutionary change is called "symbiogenesis." These mergers, long-term biological fusions beginning as symbiosis, are the engine of species evolution.
>
> (Margulis, L. and Sagan, D., 2002. *Acquiring Genomes: A theory of the origins of species.* New York: Basic Books. p. 12)

> For two different types of genomes to merge and form a new one, the organisms themselves must have a reason to come together. Reasons vary. Organism A may find B delicious, and try to swallow B. Alternatively, organism A may require the chemical form of nitrogen excreted in the waste of B. Organism A may simply bask, at first, in the shade provided by B – or A may sequester the alkaline moisture that exudes at dawn from the pores of B. These are ecological issues with many subtleties, but they underlie the transfer and eventual merger of microbial genomes to larger forms of life.
>
> (Margulis, L. and Sagan, D., 2002. *Acquiring Genomes: A theory of the origins of species.* New York: Basic Books. p. 89)

In this view natural selection is only a minor evolutionary mechanism; the central evolutionary mechanism is symbiogenesis. This is clearly concordant with the view that I have been outlining which entails that natural selection is a minor evolutionary mechanism which merely determines the branches of the evolutionary tree, the trunk of the evolutionary tree being determined by a more fundamental evolutionary mechanism. Could symbiogenesis be part of the fundamental feeling evolutionary mechanism that I have been outlining? Yes. From the perspective of the feeling universe that I have outlined we can interpret what Margulis and Sagan are saying as follows. They say that: "the organisms themselves must have a reason to come together. Reasons vary." We can agree that reasons vary but we can also add that all of these reasons are feeling-motivated. To say that "Organism A finds Organism B delicious" is a way of saying that Organism A moves to a more optimal state of feeling when it encounters Organism B, it therefore tries to swallow Organism B. To say that "organism A may require the chemical form of nitrogen excreted in the waste of B", is to say that Organism A moves to a more optimal state of feeling when it receives this excreted waste. To say that "Organism A may simply bask, at first, in the shade provided by B" is simply to say that Organism A moves to a more optimal state of feeling when it basks in the shade of Organism B. And, to say that "A may sequester the alkaline moisture that exudes at dawn from the pores of B" is to say that Organism A enters into a more optimal state of feeling when the alkaline moisture from the pores of B gets exuded onto it. Why else would Organism A act in these ways? It feels good for Organism A to be exuded upon in this way, to bask in this way, to try and swallow in this way.

The Evolutionary Processes Which Propel the Planet

So, in the realm of planetary life, we have an evolutionary mechanism which can explain how the feelings inherent in all parts of the universe bring forth new species and propel the evolution of planetary life along its inevitable trajectory towards the optimal state of feeling that is the technological 'human' form. This mechanism partly forms the trunk from which sprouts the branches of natural selection. Let us now consider the second evolutionary mechanism which determines the trunk of the evolutionary trajectory of life and non-life on Earth; this is the structure of the Solar System.

In order to fully understand the evolutionary processes and trajectories on the Earth we cannot simply look at the Earth in isolation; we also need to consider non-Earth factors. We need to see the Earth as simply one part of the Solar-Systic whole. By exploring this whole we can gain a deeper appreciation of the evolutionary trajectory of the Earth. So, the second evolutionary mechanism which determines the trunk of the evolutionary trajectory of life and non-life on Earth is the structure and (interconnected) nature of the Solar System.

The Earth is an evolving whole which is part of a larger evolving whole – the Solar System. What happens in one part of the Solar System is not irrelevant to what happens in another part. For example, the activity of the Sun affects what happens on the Earth. Indeed, we have already explored this interconnectedness through the phenomenon of non-human-induced global warming. As the Sun ages it sends a continuously increasing amount of solar radiation to the Earth and this obviously affects the evolutionary processes and trajectories which are occurring on the Earth. In other words, the trunk we have been exploring – the mechanism which is the heart of the evolutionary trajectory of life and non-life on Earth – is partially shaped by the amount of solar radiation reaching the Earth. This shouldn't surprise us; after all, in a feeling universe, solar radiation is itself a state of feeling which affects the states of feeling of the Earth when it interacts with them.

This interconnectedness operates at a deep level. It runs so deep that if one had no knowledge of the state of the Earth, and one just had knowledge concerning the age of the Sun, then one would, I believe, be able to make very accurate predictions about the state of the development of the evolutionary trajectories on the Earth. For example, given the present age of the Sun, one could predict that the Earth should have evolved to the stage where it is bringing forth advanced technology. This prediction would apply to any solar system in which an Earth-like planet existed at the same distance from an equally-aged sun to our own.

The interconnectedness of the Solar System doesn't apply only to the Sun-Earth relationship; it also applies to the relationship between the Earth and the other planets of the Solar System. This idea is not a new one; it has a long history through the astrological tradition. In support of this view I would like to draw on the extensive research of Professor Richard Tarnas. He has carried out a ground-breaking survey of the correspondences between the alignments of the outer planets of the Solar System and the evolutionary progression of human culture (*Cosmos and Psyche,* Plume, 2007). He details in depth how the events which are happening on the Earth are related to the positions of the outer planets; when certain alignments of these planets occur certain archetypal dynamics are activated on the Earth and these lead to corresponding advances in the evolution of human culture. This means that one can make future predictions about the type of events that are likely to be happening on the Earth, as Tarnas successfully does in his book, on the basis of future planetary alignments. Tarnas concludes that:

> The evidence suggests rather that the cosmos is intrinsically meaningful to and coherent with human consciousness; that the Earth is a significant focal point of this meaning, a moving center of cosmic meaning in an evolving universe, as is each individual human being; that time is not only quantitative but qualitative in character, and that different periods of time are informed by tangibly different archetypal dynamics; and, finally, that the cosmos as a living whole appears to be informed by some kind of pervasive creative intelligence – an intelligence, judging by the data, of scarcely conceivable power, complexity, and aesthetic subtlety, yet one with which human intelligence is intimately connected, and in which it can consciously participate. I believe that a widespread understanding of the potent but usually unconscious archetypal dynamics that coincide with planetary cycles and alignments, both in individual lives and in the historical process, can play a crucial role in the positive unfolding of our collective future.
>
> (*Cosmos and Psyche,* Plume, 2007, p. 489)

In using his research to make future predictions Tarnas states that:

> If we can judge by past experience, the most significant and potentially dramatic configuration on the horizon is the convergence of three planetary cycles that will produce a close *T-square* alignment of *Saturn, Uranus,* and *Pluto* during the period 2008-11. The last time that these three planets were simultaneously in hard aspect was from 1964 to early 1968, when Saturn opposed the longer Uranus-Pluto conjunction of the 1960s and when both revolutionary and reactionary impulses were intensely constellated and complexly interpenetrating in the collective psyche... The forces involved seemed to demand, as well as bring forth the possibility of, a deepened capacity for the creative resolution of intensely opposing forces – the old and the new, the past and the future, order and change, tradition and innovation, stability and freedom. A general atmosphere of power struggle is typical. Underlying tensions between established social authority and newly empowered countercultural impulses tend to be exacerbated. So also the generational tensions between old and young and the political tensions between conservative and progressive... the result is a destructive encounter between forces of revolutionary change and forces of rigid reaction
>
> (*Cosmos and Psyche,* Plume, 2007, p. 479)

This prediction concerning the type of events likely to be dominating on the Earth in the 2008-2011 period, a prediction based on the close *T-square* alignment of three of the outer planets, seems to me to have come true. The description seems to fit perfectly with the 'Arab Spring' which erupted in 2010. The 'Arab Spring' is the revolutionary wave of demonstrations and protests (both non-violent and violent), riots, and civil wars that have spread through the Arab world. The demonstrators have sought to bring down the authorities/regime, and the authorities (the established order) have typically reacted with a violent response. It is clearly of interest that the archetypes/feelings that led to such momentous change, at this particular time, were predictable on the basis of the prevailing Solar-Systic structure.

So, the trunk of the evolutionary progression of life on Earth is determined jointly by the nature of the feeling universe itself (the feeling states of the Earth yearning to move to a better state of feeling) and by the structure of the Solar System. The Solar System is structured in such a way that the positions of the outer planets – Saturn, Neptune, Pluto, Uranus, and Jupiter – mirror the events which are occurring on the Earth. These events include the propelling of life towards the human species, and the subsequent propelling of human cultural evolution towards globalised technological society. The evolutionary stage of life on Earth is also intimately connected to the ageing of the Sun.

I have claimed that the evolutionary trunk of both life on Earth and non-life on Earth is constituted by two separate evolutionary mechanisms. The first mechanism is the feeling/yearning nature of the universe; the second mechanism is the structure, and interconnected nature, of the Solar System. These two evolutionary mechanisms are deeply intertwined. For, the current feeling states of the Earth will be partly determined by the current state of the Solar-Systic Structure (which is a structure of various feelings). When there is a particular Solar-Systic Structure (a particular alignment of outer planets) then certain archetypal dynamics are activated on the Earth. We can think of certain archetypal dynamics as corresponding to a certain range of feeling states. In other words, as the outer planets circle around the Earth, there will be a corresponding change in the feeling states on the Earth. So, we can think of the evolutionary trunk of life on Earth and non-life on Earth as actually being constituted by *a single mechanism*. The feeling states of the Earth are yearning to move to a more optimal state, and these feeling states themselves are affected by the current structure of the Solar System (the current relationship between the feelings/position of the Earth and the feelings/position of the outer planets of the Solar System). It is one dynamic unfolding process.

Of course, this *single mechanism* is the *singular force* that I have conceptually divided into *the force to environmental destruction* and *the force to environmental sustainability*. When I introduced this force I stated that for the vast majority of the history of the Solar System it has been equivalent to *the force to environmental destruction;* however, that at a particular stage in the evolution of the Earth-Solar System this force split itself into two and brought forth *the force to environmental sustainability.* We can now appreciate this from a broader perspective. The Solar System is a singular entity which slowly ages through time, this ageing process is characterised, and catalysed, by changing archetypal dynamics which are determined by the prevailing Solar-Systic Structure. These changing archetypal dynamics (these changing patterns of feeling states) bring forth and propel human culture on the Earth; this bringing forth and propelling is dominated by archetypal dynamics/feelings which can be aptly characterised as *the force to environmental destruction*. The maturation of human culture leads to a changing dynamic; the maturation of human culture is simultaneously a maturation of feeling states, it is also a maturation of the (feeling states of the) evolving Solar-Systic Structure. When human culture reaches a particular stage in its evolution (the technological birthing stage), these changing archetypal dynamics give rise to an increasing number of feeling states which encourage 'sustainability' rather than 'destruction'; in this way *the force to environmental sustainability* is born. So, rather than there actually being two forces, we can more aptly think of there being a gradual maturation, a gradual change in nature, of a singular force.

We can conclude from all of this that the Solar System is one entity and that within that entity the Earth is the womb. The Earth is the womb of Solar-Systic life and the entire Solar System is urging the Earth forwards, propelling it towards a successful technological birth.

Chapter 11

Humans in the Cosmos

We have been exploring the relationship between the human species and the cosmos of which it is a part. Our objective has been to build up a broader understanding of how this relationship impinges on issues relating to global warming. We have made a lot of progress. We have considered the two different types of global warming that exist and affect the Earth. We have considered the benefits of technology for life on Earth. We have considered the nature of life and what it means to be human. We have considered the different aspects of the environmental crisis. We have considered the relationship between technology and the environmental crisis. We have considered the history of the Solar System and the nature of the universe. We have also considered the evolutionary mechanisms which propel planetary life forward towards technology.

In this chapter I would like to consider some further aspects of the human/cosmos relationship. I would like to explore how the philosophical worldview that I have outlined can have practical significance for the well-being of individual humans, for the well-being of humanity, and for the well-being of life on Earth. I would also like to further explore the relationship between humans and the non-human life-forms of the Earth, and the stories that we tell ourselves about this relationship.

Let us start with the relationship between humans and the non-human life-forms of the Earth. Within the human species there is a very widespread sense that the human species is superior to all of the non-human life-forms of the Earth. This sense reveals itself in a plethora of ways: we eat animals on a massive scale; we factory farm and slaughter animals in their millions; we perform vivisection on animals; we use animals for scientific experiments which couldn't ethically be done on humans; we poison animals; we kill animals for fun, for sport; we have a great number of religions which support the idea of 'human dominion' over the rest of life on Earth; we cage animals in zoos; we widely sell animals like commodities; we insult other humans by using phrases such as *you lowlife, you are nothing but a dirty animal!*

I believe that this sense that the human species is superior is not something that is rationalised; it is something that leads to rationalisations. Firstly, a 'sense of superiority' arises automatically in a human (due to growing up 'surrounded' by advanced human tools/technology). Secondly, a human might ponder the following question: What is it that makes the human species superior? In other words: Why isn't the human species just another species of animal? This question has been answered in a multitude of ways throughout human history. Some of the answers that humans have come up with are that humans are superior because of: soul possession, consciousness, rationality, intelligence, feeling/emotion, culture, self-awareness, language, tool use or morality.

We are living at a time when this process of rationalisation has actually led many people to doubt the validity of their 'sense of superiority'. In other words, despite having a sense that the human species is superior to all of the non-human life-forms of the Earth, many people have rationally convinced themselves that in reality the human species is not actually superior in this way. There are four main reasons why this has happened. Firstly, the legacy of Charles Darwin has led many people to conclude that the sense of human superiority is not compatible with the hard scientific facts concerning the way that evolution occurs. Secondly, the attributes that humans have typically used to rationalise/justify their sense of superiority seem increasingly implausible. It is becoming increasingly widely accepted that some non-human Earthly life-forms use tools, are rational, use language, have feelings/emotions, have morality, have their own culture, and so on. Thirdly, one impact of growing environmental awareness has been to lead to a counter-rationalisation: humans aren't superior they are destroyers, the parasites of the planet. Fourthly, a recent evolution in human thought known as postmodernism urges us to accept that there is no objective reality; there are just a plethora of differing perspectives.

The combination of these four reasons has made it intellectually fashionable to believe that the human species is not superior to the non-human life-forms of the Earth. It is extremely likely that you will have heard someone assert something along these lines: *Humans consider themselves to be a superior species, but surely the dolphins consider themselves to be a superior species, the elephants consider themselves to be a superior species, and the chimpanzees consider themselves to be a superior species.* You might even have heard something like this: *The idea that the human species is superior is nothing but egotistic anthropocentric nonsense!*

What should we make of such assertions? We should reject the second and think of it as just an emotional outburst which comes from a perspective of very limited understanding. When it comes to the first assertion we can agree that this might be true. What we are ultimately concerned with is not the 'sense of superiority' that an individual life-form has concerning the superiority of its species. What we are really interested in is another phenomenon: reality. We are interested in the question of whether in reality the human species has a position of superiority. This reality is independent of the 'sense of superiority', and it is also independent of 'human rationalisations'. The reality is the reality. The rationalizations change over time like fads; in one epoch humans are widely developing various rationalizations that support human superiority, in the next epoch they are widely developing rationalisations that undermine this superiority. As these epochs come and go the reality hasn't changed.

So, let us accept the possibility that numerous non-human Earthly life-forms either have a sense that their species is superior, or rationalise that this is so. Whilst wishing to accept this possibility, I should make it clear that I believe that only the human species actually has a 'sense of superiority'; this is because I believe that the cause of this sense is growing up 'surrounded' by the advanced tools/technology that have been brought forth by one's species.

Let us return to the four reasons that have combined to make it intellectually fashionable to believe that the human species is not superior to the non-human life-forms of the Earth. The first reason is the legacy of Charles Darwin. In *Chapter Ten* I outlined the nature of the evolutionary forces which propel the progression of both life on Earth and non-life on Earth; we saw that Darwin's theory of natural selection has a role to play in this process, but only a minor role. If one believes that Darwin had the final word on the fundamental nature of the evolutionary processes which occur on the planet, then one will find it hard to believe that the human species is the zenith of the evolutionary progression of planetary life; however, if one denies this (that the human species is the zenith) then one will have wrong beliefs. There is a strangely widespread tendency for people to think that they understand the fundamental nature of the evolutionary forces which propel planetary life, and they seem to think that they understand because they mistakenly believe that Darwin himself had such a fundamental understanding.

The second reason is the fact that the attributes that humans have typically used to rationalise/justify their sense of superiority seem increasingly implausible. This is correct; these attributes are implausible. Humans have been grasping around for a superior-making attribute for millennia without any luck. The mistake has been to think that a direct comparison between an individual human and an individual non-human (say a chimpanzee) will reveal a superior-making attribute that the human possesses. This just isn't the case. To find the source of human superiority one needs to see the entire human species as fulfilling a special role in a process, the unfolding of life on Earth as it enters the epoch of technological birthing.

The third reason is that growing environmental awareness has led to a counter-rationalisation: humans aren't superior; they are destroyers, the parasites of the planet. The entire purpose of this book is to show why this counter-rationalisation is wrong; it is based on a very limited and partial understanding. There is human destruction, but this is perfectly compatible with human superiority; indeed, destruction is required for the human species to act as the technological saviour of planetary life.

The fourth reason is that postmodernism urges us to accept that there is no objective reality; there are just a plethora of differing perspectives. The people who have been sucked into this way of thinking have seemingly failed to realise that their entire postmodern view is itself just a perspective. This view might be useful in some areas of human enquiry and endeavour, but the reality of human superiority is not something which is just one perspective among many.

So, when we delve into these four reasons we find that they don't provide strong support for the view that the human species is not superior. My objective throughout this book is to convince you that in reality the human species is superior to all of the non-human life-forms of the Earth. This is so because the human species occupies a uniquely special place in the evolutionary progression of life on Earth. The human species is that part of life on Earth that has become technological, and the bringing forth of technology, the epoch of technological birthing, is the zenith of the evolutionary progression of life on Earth. Another objective that I have is to convince you that this reality is not something that you should be worried or concerned about. You might be worried that accepting the notion of human superiority would lead to further exploitation of non-human life-forms, and of the environment. Far from this being the case, I believe that the opposite outcome is true. Fully appreciating the superiority of the human species can lead to a more harmonious relationship with the non-human life-forms of the Earth, and it can lead to a more harmonious and sustainable future. Furthermore, the philosophical worldview which this notion of human superiority is embedded in can also lead to individual humans having a more fulfilled and happy existence.

The relationship between the human species and the non-human life-forms of the Earth is a relationship that changes over time. The reason for this change is that the human species is that part of planetary life which is the bringer forth of technology, and this bringing forth is a three stage process. In the recent past, when there were just life-forms using non-advanced tools, then the human species did not exist (the human species *is* the bringer forth of technology, so, no technology=no human species). It is when these life-forms, these biological creatures, became technological, that the human species was born. The birth of the human species was not a biological event, it was a technological event. Before the human species came into existence there were still biological individuals that most people would, when they consider history, normally refer to as humans. These biological individuals first evolved at a particular moment in time and their subsequent evolution, growth in numbers, and creation of non-advanced tools, represents the first stage of the technological birthing process – the 'pre-human stage'. In this stage these organisms are preparing the groundwork for the technological birthing process, and they progressively start to impinge on the welfare of the rest of the life-forms on the Earth. However, their overall impact is still very small.

In the second stage of the technological birthing process the human species springs into existence, technology evolves and multiplies; this is the 'technological birthing' stage, or the 'human dominion' stage.

This stage involves a change for non-human planetary life, which is now being affected much more severely. In effect, non-human planetary life becomes the servant of technology. More than this, it becomes the 'willing' servant of technology; after all, planetary life has been yearning for the technological birth for millions of years because it benefits so much from such a birth. This is the stage of human exploitation of non-human planetary life. This is the stage we are still living through in which the human species exploits non-human planetary life-forms because they are widely perceived as inferior. This perception is an inevitable prerequisite for a successful technological birth. One can only dissect, explore, manipulate and utilise something if one considers it to be inferior.

In the third stage of the technological birthing process the birth has been completed and we enter the 'post-birth' stage. In this stage the epoch of human dominion is over. There is a widespread realisation that the human species is fundamentally similar to, and part of, the totality that is planetary life. There is a widespread realisation that all life-forms are deeply precious, a realisation that all life-forms are 'brothers/sisters in striving/yearning'; planetary life strives for survival and every part of it is precious, every part of it is deserving of respect. This stage is a stage of harmony and of joyful peaceful coexistence between the human species and non-human planetary life. It is also a stage in which the human species simultaneously becomes horrified at the harm that it previously caused its brothers/sisters, and appreciative of the inevitability and positive outcome of such harm.

We are currently in the second stage, the 'human dominion' stage. We see ourselves as superior and this justifies the harm that we cause to non-human planetary life. Now, of course, in reality we are superior. There is no point in denying this! But if we come to appreciate exactly *why* we are superior then this could have wonderful results. If it is widely realised that we are superior for the reasons outlined above, for the reasons developed throughout this book, then we can see our place on the Earth, our place in the Solar System, in a fundamentally new light. This new vision can naturally speed up our transition into the third stage of the technological birthing process, the stage of harmony and joy.

We are at the stage where the completion of the technological birthing process is imminent; as I have already explained, the birth itself cannot be aborted. However, we do have the ability to consciously push forwards into stage three with great haste, if we so wish. We can achieve this by firstly accepting our superiority and secondly by completing the technological birth as soon as possible (which essentially means technologically regulating the atmospheric temperature). Realising that (and why) we are superior to non-human planetary life, fully embracing this, can speed up our transition to the wonderful future in which human and non-human live peacefully, joyously, respectfully and harmoniously. Let us embrace our superiority!

We have been exploring how the philosophical worldview that I have outlined has practical significance for the well-being of humanity, and for the well-being of life on Earth. Through accepting our superiority, our special place on the planet, we can speed up the technological birthing process and gain a new positive view of human cosmic purpose. This will lead to a positive outcome for life on Earth – the rift between humanity and non-human planetary life can be more speedily closed; this will result in less harm and suffering both within humanity as a whole and within non-human planetary life. Let us now consider how the philosophical worldview that I have outlined has practical significance for the well-being of individual humans. There are two things to keep in mind:

- Individual humans are born into a particular stage of the larger unfolding Solar-Systic and planetary processes that we have been considering.

- Individual humans are born with a particular set of capacities and abilities, the nature of which is determined by the Solar-Systic structure in the period in which the individual is being forged/brought forth (we considered the importance of Solar-Systic structure for the events happening on the Earth in *Chapter Ten*); these capacities and abilities may or may not be fully activated within the life of an individual.

The former provides a backdrop, a framing, for the latter. In other words, the stage of cultural evolution into which an individual human is born provides a number of possibilities for that human to utilise their capacities and abilities. If one has the capacity and ability to be an amazing inventor then one might or might not be able to invent the wheel, the telephone, the laptop computer; it depends when one is born. One clearly cannot invent something if one is born into a stage of the unfolding process in which that thing has already been brought forth into existence!

If a human utilises their capacities and abilities through fulfilling some of the available possibilities for their deployment, then they will generate states of good feeling within themselves. A human who has the capacity and ability to be a great inventor feels good when they are inventing; a human who has the capacity and ability to be a great painter feels good when they are painting; a human who has the capacity and ability to be a great musician feels good when they are composing/playing. Some humans realise from a young age which activities make them feel good in this way, fulfilled, happy, content, 'in the flow', 'self-actualised'; these humans have their entire lives to engage in these activities. Other humans come to this realisation later in their lives and then start to act accordingly. There are some humans who never come to appreciate what their capacities and abilities are; these humans never truly feel good; they live unhappy lives; they might even be driven to suicide. Thankfully, the vast majority of humans come to realise what their capacities and abilities are and also find activities which enable them to utilise these attributes. This enables them to lead happier lives, to generate good feeling states, and to fulfil their potential to a greater or lesser extent.

I am suggesting that individual humans have particular sets of capacities and abilities and that if a particular human is living in accordance with their capacities/abilities that they will feel good/content/happy. You will probably agree that this seems to be so; it is not an outlandish assertion! I am also suggesting that the majority of people come to realise what their capacities/abilities are at some stage in their lives and act in accordance with them. Again, I doubt that you think that this is a controversial suggestion. When we 'scale up' and zoom out we can view individual humans as part of the unfolding of human culture on the Earth. When we do this we can see that the trajectory of human cultural evolution is propelled forwards by the individual humans that are alive at a particular moment in time and who are engaged in some of the suitable activities that are available for them to do.

At any moment in time there will be a certain number of activities which an individual can partake in in order to utilise their capacities/abilities. For example, if one is an inventor and one is born into the stage of the unfolding process when the wheel has already been invented then one cannot invent the wheel. However, one could possibly invent the bicycle, the wheelbarrow, the skateboard, or the rollerblade, if these things have not already been brought forth into existence. Through this example I am trying to get you to appreciate that at any moment in time there are a *certain range of activities* which humans with certain sets of capacities/abilities can do in order to feel good/fulfilled. At any moment in time there are also a *certain number of humans* which have the capacity/ability to carry out a certain activity. In other words, before the wheel was invented there were lots of humans who had the capacity/ability to invent the wheel.

The realization that these two groups exist (humans with particular capacities/abilities, and available activities for the exercise of these capacities/abilities) is linked to the question of individual human purpose. Many humans at some stage in their lives ask themselves: *What is the purpose of my life?* When this question is asked a great many people are looking for a specific answer. For example, the purpose of my life is to become a member of parliament; the purpose of my life is to climb Mount Everest; the purpose of my life is to invent a time machine; the purpose of my life is to develop a new theory of everything. All of these things might actually be the purpose of someone's life. A particular individual might feel so exhilarated about politics, so passionate, that their entire body might tingle with delight when they are in the House of Commons. It would make sense to say that the purpose of this individual's life is to become a member of parliament.

However, I think it is best to approach the question of individual human purpose in a general way. I don't think that, for example, there was a particular individual human who was born on the Earth with the purpose of inventing the wheel. This seems wildly implausible to me. I think the best way of thinking about individual human purpose is as follows:

Question: *What is the purpose of my life?*

Answer: It is to do that which makes me feel happy/ content /exhilarated /self-actualized.

When we think of individual human purpose in this way it is clear that it has two elements. Firstly, the activities which are available to do given the stage of the cultural evolutionary process. Secondly, the capacities/abilities of a particular human the exercise of which leads to happiness/contentment/exhilaration/actualization. There needs to be a match between these two elements. Viewing individual human purpose in this way is useful because it fits the individual human into the larger evolutionary unfolding of human culture/planetary life.

When we zoom out and consider the entire evolutionary trajectory of human culture, from hunter gatherer to globalised technological society, we can ponder a particular moment in time, a particular part of that process. We can see that at this moment in time there were a certain number of activities which needed to be carried out if human culture was to reach globalised technological society. We can, in effect, map out all of the things that needed to happen in the future from this past point. The wheel needed to be invented, money needed to be invented, capitalism needed to be invented, the airplane needed to be invented, the computer needed to be invented, the television needed to be invented, the United Nations needed to be invented. The complete list would be exceptionally long! We can think of these individual required occurrences as 'mini purposes' which are stepping stones along the evolutionary unfolding path to globalised technological society. From the perspective of this unfolding process there is another answer to our question:

Question: *What is the purpose of my life?*

Answer: It is to fulfil 'mini purposes' in order to aid the evolutionary unfolding of planetary life.

These two answers amount to exactly the same thing, they just come from a different perspective. One can answer from the perspective of the individual, or one can answer from the perspective of the whole of which the individual is a part. The capacities/abilities of humans do not arise from within in independence from that which is without. Indeed, we can make perfect sense of the idea that the entire Solar System is involved in the forging of the capacities/abilities of a particular individual human. In *Chapter Ten* we briefly explored the astrological research of Professor Richard Tarnas which points to this very conclusion. When the Solar-Systic structure is in a certain arrangement, then individuals with a certain type of capacities/abilities will be born. When the Solar-Systic structure is in a different arrangement, then individuals with different capacities/abilities will be born. In this way the Solar System ensures that it brings forth all of the capacities/abilities that it requires for a successful progression of human culture to globalised technological society. In other words, this ensures that the technological regulation of the Earth's atmospheric temperature will occur.

So, the capacities/abilities of individual humans have been endowed to them by planetary life/the Solar System so that, in exercising these capacities/abilities, the humans concerned (and thus planetary life, and the Solar System, of which that human is a part) will move to a more optimal state of feeling. In acting in this way humans give a trajectory to human culture through the slow but steady and inevitable fulfilment of 'mini purposes'. This process takes the evolutionary progression of planetary life all the way to its technological zenith.

In short, to maximise one's well-being one simply needs to utilise one's capacities/abilities as fully as one can. In doing this one will not only be creating an exhilarating/fulfilling life for oneself, one will also be performing a service for life on Earth through speeding up the evolutionary unfolding of human culture.

Chapter 12

The Interplay between Technology and Spirituality

It is obvious that the human species is that part of life on Earth that has become technological. One does not need to be particularly insightful to be able to realise that this is so; all one needs is the ability to observe the human and non-human life-forms of the Earth. The contentious issue is what we should make of this obvious fact. It is obvious to me that this obvious fact is of crucial importance. It is what makes the human species the human species. It is what carves an essential distinction between the human species and all of the other life-forms of the Earth. It is what 'raises' the human species up to a position of superiority over the rest of life on Earth.

One can obviously take the obvious fact to a different conclusion. One could be nonplussed by the obvious fact. One could think: *So what? Humans are technological, birds are great flyers, cheetahs are great runners, whales are great swimmers, and so on.* One could also take the obvious fact to a negative conclusion: *The human species is that part of life on Earth which has become technological. How terrible! What destruction we are doing! If only our technology could be got rid of. If only we would go extinct; then the planet could return to a state of non-technological harmony.* The entire purpose of this book is to persuade you that these two views – the nonplussed and negative views – are wholly misplaced.

The human species has a special place on the planet and it is superior to the rest of planetary life because it is the bringer forth of technology. There is no *other reason* why the human species is superior. Take away our role as the bringers forth of technology and we become a wholly non-superior form of life; we become just 'individual' interconnected life-forms (just part of the uniform oneness of planetary life). This is blatantly obvious to me. However, I realise that you might be uncomfortable with this conclusion. One possible reason for this uncomfort is the belief that humans are ultimately superior because they are spiritual beings. If one believes this, then spirituality will trump technology in one's view of the human/non-human relationship. In other words, one could believe that technology makes the human species superior, be nonplussed by technology, or believe that it is a negative phenomenon, and whichever of these views one adopted would be of no real importance. If one believes that spirituality raises the human species to a position of superiority, then it makes no difference to one's view what is going on in the realm of technology.

How does spirituality fit into the wider philosophical worldview I am outlining in this book? In order to address this question I will initially describe the technological and spiritual processes which are unfolding on the planet. Let us start with technology. The evolution of technology is a long process. The first life-forms to evolve on the Earth (single-celled life-forms) utilised the resources which existed in their surroundings through metabolising, and thereby modified their external surroundings. We can certainly think of this as very primitive 'tool use'. As life progressed, multicellular organisms evolved which had a greater ability to modify their surroundings. Some of these organisms became what are standardly referred to as 'tool users'; for example, chimpanzees using sticks to withdraw honey from beehives and for dipping ants.

A standard definition of a tool is *something that is held in the hand, foot or mouth and used so as to enable the operator to attain an immediate goal.* This use of tools is an advancement from the very primitive 'tool use' that came before. Of course, many people would say that what came before wasn't really tool use, it was just the utilisation and modification of the surroundings; however, the definition of 'tool use' that these people standardly use is very arbitrary. Why shouldn't one be able to use a tool in a multitude of ways? For example, why shouldn't holding something in one's armpit classify as tool use, or attaching something to one's head. We can also question the validity of the distinction between 'tool' and 'body'. Why shouldn't using one's body to utilise the resources that exist in one's environment to attain an immediate goal (which is what the first single-celled life-forms to evolve on the Earth did through metabolising) also be thought of as tool use? It is the act of utilising and modifying resources that is fundamental; the distinction between body and tool is much more illusory. For example, imagine a human who has lost their hand and has had it replaced with a mechanical contraption. When this human uses their mechanical hand to write, the pen is a tool that they are using, but is their mechanical hand also a tool, or is it just the part of the body that is using the tool? I think we need to think of the entire evolution of life on Earth as a singular increasingly sophisticated process, a process of greater control and manipulation over that which is not one; a process of increasingly sophisticated tool use. Whether or not we decide to call the initial stages of this *singular increasingly sophisticated process* 'tool use' is of little importance.

After a prolonged period of 'tool use' (as standardly defined) by both non-human and human life-forms, planetary life moved to the next level; via the human species it became technological. Technology evolved and at first it very gradually complexified, but when this complexification reached a certain stage of development a tipping point was passed. After the tipping point technology 'exploded'. It had an almost exponentially increasing self-generated momentum, it pervaded the planet, and it burst forth in glorious fashion in a multitude of ways. We are currently living in the midst of this epoch – the epoch of the technological explosion.

Let us turn to the path of spiritual unfolding on the planet. Were the first life-forms to evolve on the Earth spiritual beings? I have claimed that these first life-forms were primitive 'tool users', and we can make a similar case with regards to spirituality. After all, I believe the central tenet of spirituality is to attempt to increasingly live in accordance with the inner wisdom, the inner feelings, that lie within one. I should explain exactly what I have in mind when I talk of 'inner feelings'.

Let us start by exploring the 'inner feelings' of a particular human. The inner feelings of this human are the feeling, the felt sense, the inner felt knowledge, the inner wisdom, that what they are currently doing is right or wrong, or that what they are currently experiencing is right or wrong. Some of the inner feelings that exist in this human will exist in the overwhelming majority of humans. For example, when this human starts to walk towards a blazing bonfire then, as they get very close to it, they will start to experience feelings within them that indicate that it is a good idea to stop walking. These inner feelings might simply be 'intense discomfort' or 'mild pain'. Another example is the feeling that this human has when they witness the accidental death of a group of schoolchildren, or a child being savagely attacked by a dog; the overwhelming majority of humans will experience bad/sad/sickening/repulsive feelings in response to such events.

Other inner feelings in this particular human will be more unique because they will relate to their own particular individual capacities/abilities/capabilities. If this human has the potential to be a great teacher, but expends their energy cleaning rather than in the domain of teaching, then there is a clear mismatch; their capacities/abilities are not in accord with their life actions, and the result is deleterious/painful/uncomfortable inner feelings. If this human sets out on the road to teaching, then the resulting match between their capacities/abilities and their life actions will lead to blissful/harmonious/happy/exhilarating inner feelings. We should clearly seek to live in accordance with our inner feelings/inner wisdom as fully as possible.

What I am saying here applies to the overwhelming majority of humans, at the very least. For this overwhelming majority, living more fully in accordance with their inner feelings leads to a more fulfilling, compassionate, ethical, and spiritual existence. I am at least open to the possibility that there are exceptions to the rule. By this I mean that it is plausible that a very small minority of humans genuinely feel a great sense of joy in the face of tragedy, they feel joy at the death of the schoolchildren, and they feel joy when they carry out heinous crimes. If this is genuinely possible, then these humans clearly shouldn't act in accordance with their inner feelings; they should be restrained and locked up. I am not myself convinced that this is possible; if someone *appears* to be joyful in the face of tragedy, if they can joyously carry out heinous crimes, then this seems to be a sign that they are completely disconnected from their inner wisdom/inner feelings.

I believe that the entire universe has interiority (inner feeling states) and that these states direct and propel the evolution of the universe. So, the first (single-celled) life-forms to evolve on the Earth had feeling states. We can think of these feeling states in a more simple way than when it comes to the existence of these states in humans. These simple life-forms were simply moving in response to their inner feeling states, and this is what caused their evolution into more complex multicellular organisms. In other words, for these simple life-forms there was an *automatic* movement 'towards' what felt good and 'away' from what felt bad. Because these first life-forms were living fully in accordance with their inner feelings, in one sense they can be thought of as 'living a spiritual life'.

From these simple beginnings there was a spiritual unfolding as life-forms complexified. The inner feeling states that existed in the first life-forms to inhabit the Earth are analogous to the inner feeling states that exist in humans today. However, we seem to be fundamentally different to these simple life-forms. We seem to have a certain amount of choice concerning whether or not to live in accordance with our inner feeling states. Indeed, the existence of such a choice, for many, will be a prerequisite for spirituality. The ancient life-forms were automatically 'spiritual' beings because they lacked the ability to reason themselves to alternative actions. At some point in the evolution of life the ability of life to pursue alternative courses of action arose; this possibility might be thought of as 'free will', or 'rational ability' or 'thought' or 'reflective awareness'.

The evolution of this possibility for alternative courses of action meant that some life-forms, such as humans, might not be living in accordance with their inner feelings/inner wisdom. I should stress that this isn't an either/or situation; living in accordance with one's inner feelings is a matter of degree. I believe that the vast majority of humans have always lived to a fairly high degree in accordance with their inner feelings; for, to not do so leads to mental and physical illness and to numerous kinds of suffering. Doing so is what has propelled human culture from hunter-gatherer to globalised technological society. However, there is almost always scope for humans to live more in accordance with their inner feelings than they currently do.

The vast majority of humans live to a fairly high degree in accordance with their inner feelings without any conscious effort to be spiritual beings. The realisation that living more in accordance with one's inner feelings is a good thing to do, and the consequent pursuit of this objective, results in what is known as a spiritual journey, or a path of spiritual development or spiritual practice. A spiritual journey typically includes meditation, which is a way of connecting at a deeper level to one's inner feelings. Meditation and mindfulness are concerned with attempting to be present with one's currently existing inner feelings, so that one is not 'taken away' from these feelings by thoughts concerning the past and the present.

The epoch of the technological explosion, which I outlined a little earlier in this chapter, seems to be simultaneously an epoch in which humans have become increasingly preoccupied with the external as opposed to the internal. In other words, it seems to be an epoch in which the spiritual journey has largely been neglected by the human species. However, this surely wasn't a new phenomenon in the era of the technological explosion. In other words, for almost its entire history the vast majority of the human species seems to have been more external focused rather than spiritual journey focused. The epoch of the technological explosion simply exacerbated this pre-existing tendency and took it to its extreme. When you have reached the extreme there is only one way to go, and that is to the less extreme. Whilst we are still living in the epoch of the technological explosion we are also gradually moving into the epoch where there will be a spiritual explosion. The widespread switch from external states to inner feeling states will result in an increasingly vast proportion of the human species embarking on spiritual journeys. This is a slow and drawn out process, but it is the future.

So, one can think of the 'technological explosion' as a prerequisite for the 'spiritual explosion'. It is as if once we have fulfilled our purpose as a species, once we have brought forth the technology that life requires, that then we will be rewarded. We have suffered immensely in order to fulfil our purpose, but a glorious harmonious future awaits; it is a future of increasing spiritual fulfilment.

I would expect such interplay between technology and spirituality to exist on any life-bearing planet. So, if we were to encounter a highly spiritually evolved species of alien life-form, then there is one thing that we could be fairly certain of: *these alien life-forms come from a planet which is highly technologically advanced.* Their planet has gone through a technological explosion which was followed by a spiritual explosion.

It wouldn't be correct to say that spiritual development is a side-effect of technology. Yet, technology lays the foundations which enable spirituality to bloom. This is the reality of what occurs and unfolds on an evolving and ageing life-bearing planet. This whole process of intertwining between the evolution of technology and the evolution of spirituality – a very gradual evolution of both, leading to a technological explosion which brings forth a spiritual explosion – can be thought of as leading to the self-actualisation of the human species.

When the human species first evolved it was in no state to become spiritually advanced/self-actualised. It first needed to meet its basic survival needs, and then it needed to attain a certain level of comfort and security through increasing (technological) control of its surroundings. The drive to achieve this comfort and security arose from the feelings and motivations within humans. In becoming technological the human species was simultaneously enhancing its own comfort/security and achieving the purpose for which it was brought forth into existence, to be the technological saviour of life on Earth. When the human species has fulfilled its purpose, when the technological birthing process is complete, then the widespread explicit human spiritual journey can begin. The technological explosion enables the spiritual self-actualisation of the human species.

The capacity for spiritual development and spiritual self-actualisation does not in itself make the human species superior to the non-human life-forms of the Earth. It is the unique ability of the human species to bring forth technology which makes the human species superior. Only the human species can save life on Earth from the danger of non-human-induced global warming. Only the human species can technologically regulate the Earth's atmospheric temperature. It is solely this ability which makes the human species the zenith of the evolutionary progression of life on Earth.

Let us return to the crossroads. I hope that it is now clear to you which of the paths we need to tread. Let us joyously skip and jump along Path 2 and then continue onto Path 3.

PART 2: DIALOGUE

Objector: It is perhaps worth going back to first principles: what is causing the global warming problem? I think the science is already in; it is the burning of fossil fuels and human activity. In my opinion, naysayers tend to ply their misrepresentations of the science for personal fiscal advantage. Therefore, the most valid intervention, from my point of view, is to rein in the activity that is causing the problem.

NPC: Your view is a very common one. The problem with what you say is that it is very limited in scope; by this I mean that you are identifying a genuine problem but missing the real fundamental underlying problem. The tip of the iceberg might be a problem, but the real problem is the overwhelming mass of the iceberg which exists below the surface. You are concentrating on the tip; I am trying to show you the entire iceberg. That is to say, you are focusing solely on the science of global warming and you are unaware of the unquestioned underlying assumptions. The overwhelming mass of the iceberg is the philosophy of global warming.

Objector: I don't think that we hubristic humans have the right to start engineering a climate system we are only now beginning to partially understand. Perhaps I'm wary of giving that much control to a mere human when nature can deal with it more effectively; the Earth system is quite capable of fixing itself if humans would desist from damaging activities.

NPC: Your view is grounded in a sharp dichotomy between humanity and nature. You believe that humanity is the destroyer/damager of a self-fixing non-human nature. The reality is that humanity is nature and that human activities are part of self-fixing nature. In other words, nature brought forth human technology in order to fix the Earth's climate through geoengineering. In the absence of the human species the "Earth system" is not capable of fixing itself.

Objector: Geoengineering! Were we such intelligent engineers, we would have foreseen the problem of global warming early on and chosen to develop our societies in ways that would have avoided the problem to begin with.

NPC: There is no link between "engineering ability" and the gift of "fantastical foresight". You'll need a better argument than this.

Objector: People who actually care about life on Earth are not wrapped up in narcissistic fantasies of human conquest; they are against geoengineering.

NPC: This is wholly unhelpful. Lots of people with very diverse views about what course of action we should take actually care about life on Earth. The real issue is not caring; it is the realisation that the survival of complex life on Earth (and ultimately all life) requires the geoengineering of the temperature of the atmosphere. A planet can be wholly populated by individuals who *care about life on that planet,* yet because these individuals lack this realisation their actions can lead to that life becoming annihilated.

Objector: You outline the two paths that humanity faces and you claim that if we choose the wrong path there will be adverse consequences. Yet, you also claim that it is inevitable that we will tread the right path (the path to technological control of the atmospheric temperature). Isn't this inconsistent?

NPC: Let me explain why these two things aren't inconsistent. You are right that I believe that it is inevitable that we will tread the right path. We will tread Path 2 and ultimately Path 3. This is inevitable because the forces which are propelling the evolution of the planet, and the evolution of that part of the planet that is human culture, have already taken us to the point where our future survival requires us to tread the right path. The treading of the right path is just another inevitable stage in the evolutionary progression of life on Earth, just like the evolution of the human species was a previous inevitable stage. Yet, we still have a certain amount of choice with regards to this situation.

Imagine that a typical unarmed human is walking through a park when suddenly they become surrounded by a group of big-muscled humans carrying guns. One of the group says: "Give us your wallet". There is nobody else in the park that could intervene in this situation. I think it is fair to say that it is inevitable that the group of gun-wielding humans will acquire the wallet. Yet, the unfortunate human who is about to lose their wallet still has a choice. They can fully understand the predicament that they are in and willingly hand over their wallet. Alternatively, they could start screaming and shouting, and refuse to hand over their wallet. If they choose the second option they will still lose their wallet but they will also get beaten up and/or shot!

In other words, there is often choice within inevitability. If we initially choose Path 1, then at a later date are forced to tread Path 2 reluctantly, kicking and screaming, then like the foolish human in our wallet scenario, we will bring upon ourselves similarly undesirable outcomes (such as the preventable flooding of island states and cities, the mass creation of environmental refugees, extreme climate change events). But if we tread Path 2 consciously, willingly and proactively then the outcome will be more desirable for us; we might still get our wallet taken (face some adverse environmental impacts) but at least we won't get beaten-up and shot.

Objector: You do appear to be genuinely concerned about those who are urging us to tread what you claim to be the wrong path, the emissions reduction path to death and destruction. Given that we have some choice, genuine concern that we could actually tread the wrong path, and thereby go extinct, seems to be warranted. If you aren't genuinely concerned about this why are you bothering to write this book?

NPC: You make an interesting point. I am clearly not writing from the 'outside' in a way that provides a wholly non-causal commentary on independent and isolated events. In other words, the very writing of this book might affect what happens; it might affect which path we tread. It could be a real possibility that we tread the 'wrong' path in the absence of this book. However, the very writing of this book could itself be an inevitably stage of the evolutionary progression of life on Earth. This book could be a factor which helps put us on the 'right' path (Path 2/3). In accordance with the philosophy of 'mini-purposes' and 'individual human purpose' that I have outlined in this book, this book could be a mini purpose, and if it wasn't written by me, then a very similar book would have been written by someone else. If this is so, then genuine concern expressed in this book is compatible with an underlying lack of concern.

Objector: Somebody told me that complex life on Earth will continue for 4 billion more years.

NPC: This is possible, if we skilfully geoengineer the temperature of the atmosphere. However, I take your objection to be that complex life on Earth will continue for 4 billion more years without the geoengineering of the temperature of the atmosphere. This is wrong. You need to return to the start of the book. Alternatively, you can do some reading around Gaia Theory and Planetary Astrobiology.

Objector: We like to think of ourselves as being special. However, natural history shows that not only are we not special (our genome is undeveloped), we are also not as smart as we like to think we are.

NPC: I agree that we like to think of ourselves as a special species and that in this thinking (the reasons that we typically think make us special) we are wrong. It is a completely different question whether we are actually special (regardless of what we think of ourselves). My claim of 'specialness' has nothing to do with genomes; it is to do with 'position in life'/'position in the evolutionary process'. We are special because we are that part of planetary life which has become technological.

Objector: I believe that all species and all cosmological processes are in a state of becoming the next stage of creative evolution, creating the next opportunity for the next cosmological leap. Why should I believe that the human species is the endpoint of planetary evolution?

NPC: To see that the human species is the zenith of the evolutionary progression of life on any life-bearing planet one needs to have an adequate conceptualisation of what the human species is. For me, it is clear that the human species is the bringer forth of technology; which means that it is also that part of a planet which considers itself to be not natural (this is a requirement for developing technology). In other words, the zenith of the evolutionary progression of life is to bring forth technology to ensure the continuation of that life. Once this zenith has been reached evolution will continue; species will go extinct and new species will evolve. To appreciate that there is a zenith to the evolutionary progression of life on a planet isn't to believe that evolution will simply stop when the zenith has been reached. Indeed, it is the attainment of the zenith that enables evolution to continue.

Objector: Is the purpose of human life 'geoengineering'? No. From the perspective of modern evolutionary theory, life, including human life, is accidental and without purpose – of any kind. 'Life' (whatever that may be) just is and happens. Any presumed human 'purpose' is purely arbitrary and is a social construction, it can be argued for but it cannot be established as THE, or even A, purpose.

NPC: The view that human life is accidental and without purpose is itself a social construction. Modern evolutionary theory is not wholly false, but it is a woefully incomplete picture of the way that the universe, and the part of the universe that is life, evolves through time. What we really need is a more comprehensive view of the plethora of forces that are involved in the evolutionary process. You are right that establishing that the human species has a purpose is not easy, but that does not mean that such a claim is not true.

Objector: The "plethora of forces that are involved in the evolutionary process". What are these forces?

NPC: The entire universe is involved in an evolutionary process. This means that there is evolution in the non-living universe. Evolution in the non-living universe is determined by a range of factors. These factors are also at work in the evolution of the living part of the universe; it is just that the evolution of the living contains additional factors. The fundamental flaw in modern evolutionary theory is that it solely focuses on these additional factors at the exclusion of the more fundamental factors. These fundamental factors, the factors which determine the evolution of the non-living, and form the 'trunk' of the evolution of the living are: the interiority of the universe (feeling states) and the structure and interconnected nature of the Solar System.

Objector: Humans are obviously a destructive force on the planet. Just look around you, there is human destruction everywhere. We are on the brink of initiating another mass extinction of life on the planet. How can you possibly believe that we are the saviours of life?

NPC: The history of human thought concerning how humans relate to the wider cosmos has been pervaded by two extremes – either humans have a uniquely special and joyous place in the cosmos, or humans exist in a 'fallen' state (as in the Garden of Eden interpretation of the human condition). These two extremes have their own contemporary incarnations relating to the environmental crisis. At one extreme, humanity is the saviour of life on Earth (this would be a Noah's Ark 'technological' interpretation of the place of humanity in an evolving cosmos); at the other extreme, humans are the destroyers of life on Earth. This long-standing dichotomy of 'two extremes' can be transcended. Due to being the bringers forth of technology, humans are the destroyers of some of the life-forms of the Earth. Yet, simultaneously, humans are ultimately, and most fundamentally, the saviours of life on Earth. Far from the simple either/or dichotomy, the reality of the situation is both/and. Humans destroy in order to save.

Objector: Your argument that the continued displacement of other species by humans helps to ensure life's continuation seems highly implausible.

NPC: I think you slightly misunderstand my view. I believe that the relationship between the human species and non-human planetary life changes through time. My view is that the probability of life's continuation is enhanced by technology in various domains; furthermore, the continuation of complex life in the near future requires technology to regulate the atmospheric temperature. The development of technology *in the past* has required the displacement of species and the transformation of habitats. As humans are fundamentally kin/brothers with all life on Earth this is effectively self-harm (not good, indeed appalling, but necessary).

I don't advocate the *continued* displacement of species. Indeed, I think so much damage has been done in order to create technology for the benefit of planetary life that we now need to maximise our efforts to preserve as much wildlife/wilderness/varieties of life-forms as we can.

The *past* displacement/self-harm, and the technological gifts it has provided, enables humans to protect complex life on Earth from extinction by technologically maintaining the GMST. In the immediate future the GMST faces a double threat to spiralling upwards to a level which cannot support complex life-forms. Firstly, there is the continuing weakening of the planetary homeostatic regulatory capacity due to non-human causes. Secondly, there is the additional perturbation to this regulatory capacity which has been caused by the human release of masses of carbon from its underground storage area and its movement to a temporary home in the thermohaline circulation. Our current societal focus on the second threat, and the growing realisation of the need for geoengineering the GMST, can simultaneously solve both threats. In other words, humans have the ability to technologically save the life that has arisen on the Earth.

The future does not need to be like the past. Our population size escalated as we became technological and scientific, as did our resource use per capita; this resulted in widespread destruction of non-human life-forms and their habitats. In the future we can protect remaining habitats, value all our fellow life-forms, and save our fellow life-forms through geoengineering the GMST. These things are all jointly possible.

Objector: There are two possibilities for the future. We can work for a future in which we share the world with other species, or, we can live in a world ever more dominated by humans (and thereby enter the Anthropocene = The Age of Man). I am for sharing with other species, whereas you seem to favour the Anthropocene.

NPC: I think you have set up a false dichotomy here. I know that there are various conceptions of the Anthropocene as the next stage in the evolution of the Earth. Sometimes people use the term to refer to an increasing human dominance of the planet, and sometimes the term specifically refers to a coming epoch in which humans influence the atmosphere of the planet.

However one defines the Anthropocene, this isn't an either/or scenario as you seem to believe. A human regulated atmosphere and a human dominated planet, is perfectly compatible with "a future in which we share the world with other species". Technologically regulating the atmospheric temperature, sharing with other species, protecting habitats, treating every single life-form as exceptionally precious and valuable, all of these things are perfectly compatible. I am for all of them.

Objector: You place humans at the peak of creation, when the reality is that this is not the case; humans exist in a relationship of interdependence with all that is.

NPC: This is a simple false dichotomy. It makes perfect sense to be at the peak of an interdependent *planetary life/planet/solar system*.

Objector: The human species needs to give up its compulsion to dominate and control.

NPC: No it doesn't. It needs to accept that it is the zenith of the evolutionary progression of life on Earth; it needs to accept that its purpose is to technologically control the temperature of the Earth's atmosphere; it needs to accept that it is the saviour of life on Earth. If your 'giving up' wish came true, then we would have given up on life on Earth and the future would be dreadfully bleak.

Objector: You seem to believe that humans are the only valuable life-form on the planet!

NPC: No. I believe that all of life is exceptionally valuable in itself; in other words, the value of a life-form has nothing to do with human valuation and interests. The very fact that a life-form exists means that a precious thing exists. However, I believe that there are different levels of value within life-forms. I believe that sensations and feelings pervade the universe, existing in both the living and the non-living. In the realm of the living are plants which contain feelings but do not contain any awareness of these feelings. There is a slightly fuzzy boundary between plants and animals, and there may be a few exceptions to the rule, but animals can be thought of as life-forms which have awareness of their feelings. This awareness means that animals can suffer, whereas plants cannot suffer. For me, this means that animals are more valuable life-forms than plants; they are more deserving of our respect than are plants. Within the realm of animals there is a division between the human species and all other animals on the Earth; the human species is the most valuable life-form on the Earth. The reason for this is that the human species is that part of life on Earth which has become technological, and therefore it is the only life-form which can save the totality that is life on Earth from extinction.

Some people believe that if one accepts that all life is valuable in itself then it immediately follows that the human species should rein in its involvement with the Earth, that it should reduce its population size and its resource use; that humans should use less technology, stop modifying habitats and 'leave things to nature'. This *does not* immediately follow. Having respect for all life-forms, valuing all life-forms, does not in itself lead straight to a particular conclusion concerning the appropriate way that the human species should interact with these organisms, or with the wider non-human Earth.

Objector: You claim that the human species is a special part of the planet, the peak of creation, because it is the bringer forth of technology. This seems wrong to me. I believe that the human species is special because humans are conscious and spiritually evolving beings.

NPC: A proper understanding of our place on the planet requires an exploration of the relationship between humanity, technology and spirituality. It is obvious that the human species is that part of the evolving Earth that has become technological, and it also seems to me that the human species is a part of the evolving Earth that is spiritually evolving. I wouldn't deny that the human species contains spiritually evolving beings; however, I am convinced that this isn't the seat of human specialness. Life on Earth didn't bring forth the human species so that it could spiritually evolve; it brought forth the human species because it needed a technological saviour.

Objector: When you say that "It is obvious that the human species is that part of the evolving Earth that has become technological, and it also seems to me that the human species is a part of the evolving Earth that is spiritually evolving" this could easily be interpreted as meaning that humans are the only ones to be either technological or spiritual and that is certainly not obvious to me.

The way I see it, the use of technology is on a spectrum. Humans have developed it to an amazing degree but lots of other species use technology to some extent – crows, for example, are brilliant at devising new tools for achieving their aims. And we have absolutely no idea whether other species are 'spiritual' or not, since we can't communicate with them well enough to find out. So to claim sole ownership of it would be just another kind of hubris.

It seems to me that the whole concept of human specialness has proven so problematical that I don't think we should be claiming any kind of special status for our species any more. We do have certain skill sets that many other species don't have, thanks to our opposable thumbs and our highly developed left cerebral cortices. And we need to use these to clean up the messes we have already made and find more co-operative ways to live. What we need to do, I believe, is to apprentice ourselves to Nature in an attitude of humility and openness; this will enable our planet's best interests to be served. Permaculture is a very good example of this philosophy in action.

NPC: I will explain what I mean a little more thoroughly, because I believe that we agree more than you might realise. There is indeed a spectrum of modification. At the lower end of the spectrum 'simple' life-forms, such as microbes, modify their surroundings through metabolising. As we move up the spectrum, some animals, such as your crows, develop enhanced modification abilities which humans typically refer to as 'tool use'. As we move up the spectrum even further we move from 'tool use' to 'technology'. I am using the word 'technology' not to refer to basic tool use such as that performed by chimpanzees, crows and a plethora of other species; crows are not a technological species. 'Technology' refers solely to the top of the modification spectrum. The human species is the only part of the evolving Earth that is technological.

I wholeheartedly agree with you that: "we have absolutely no idea whether other species are 'spiritual' or not". This reinforces my belief that humans are 'separated' from the other life-forms of the Earth through being the bringers forth of technology. It is obvious that humans are the only technological life-forms on the Earth; it is not obvious that humans are separated from the other life-forms of the Earth in any other significant way – not even through being spiritual.

So, to our seeming disagreement; you say: "I don't think we should be claiming any kind of special status for our species any more". It seems obvious to me that the human species has a very special status (whether we claim it or not) in virtue of being that part of life on Earth that has become technological. You refer to our "planet's best interests" and how they are served. It is obvious to me that the planet's best interests require the utilisation of human technology in various domains. From what you say I really don't think that you appreciate what the "planet's best interests" are.

Objector: OK, point taken about tool use. So yes, by that definition we are certainly the only technological life-form on the Earth. Let me hasten to point out that I appreciate technology. It enables me to communicate effectively with others around the globe in ways I never could have done before. Technology itself is neutral. Like fire, it has the ability to create, to maintain or to destroy. But a lot of the technology we are now enjoying has come at a huge cost, i.e. the eco-destruction caused by the extraction of raw materials to build technology, the pollution and climate change resulting from its use and the human deskilling that inevitably comes from over-dependence on machines.

Then there's our seeming inability to use it wisely. Our cleverness has outrun our consciousness. For me, therefore, raising consciousness is our primary task. So I am by no means a Luddite (and I certainly favour large amounts of money being invested in appropriate technology for wind, tidal and solar power etc.) but I believe that we need to deploy it with extraordinary amounts of wisdom. I'm convinced that the best results will come from a skilful blend of the new and the old.

My wariness arises from the fact that most proponents of technological solutions to environmental problems have a vested interest in touting solutions that will actually worsen those problems because they ignore age-old wisdom. Like Monsanto trying to convince us that genetically modified crops will feed the world when in fact organic farms have better yields and are better for the Earth. So, when you talk about deploying technology to help create a sustainable world what specific applications do you have in mind?

NPC: You make a lot of interesting points relating to technology and its relationship to humanity. However, it seems to me that your assessment of technology is grounded in a 'human perspective'. You say that technology is useful because of enhanced communication possibilities for humans, technology has a huge cost because of human deskilling, humans are unable to use technology wisely, and there are vested interests in human organisations (Monsanto). Let me offer a view of technology which comes from the *Earth's perspective* (rather than the *human perspective*).

Imagine two different planets which have life on them: On Planet T (Technology) life has evolved and it has moved up the modification spectrum to the point at which the technological genie has been released from its bottle. On Planet NT (No Technology) life has evolved but the modification spectrum has yet to reach this point. Both T and NT contain a plethora of diverse and wonderful life-forms; we can assume that it will be a great shame for the cosmos if either T or NT were to return to a state of lifelessness. The question is: Is the life that has arisen on T in a better position to survive than the life on NT?

In other words: Is technology in the interests of life? And, assuming that life is in the interests of the cosmos, this equates to: Is technology in the interests of the cosmos? (These are very different issues to those of human communication, human deskilling, Monsanto, etc.) Which technologies are in the interests of life? These are not primarily the technologies that you envision as "a skilful blend of the new and the old... wind, tidal and solar power". The life on T would not be in a better cosmic position to survive than the life on NT in virtue of having tidal power. Let me outline three scenarios in which life on T survives because of its technology, whereas life on NT goes out of existence (NT returns to a lifeless state) because it lacked technology.

Scenario One: A massive asteroid is on a collision course with T/NT. Life on T is able to defend itself by utilising the technology which it has brought forth; the asteroid gets knocked off its collision course or destroyed. Life on NT has no such option and gets annihilated.

Scenario Two: In the future, due to the evolving nature of the cosmos, the Sun expires. By the time that this happens life on T has become so technologically advanced that it has already moved to planets in other solar systems. The life that has arisen on T is thriving! What of the life on poor old NT? It has been annihilated.

Scenario Three: The homeostatic regulatory ability of life on T/NT is weakening (that life on a planet, in tandem with its host planet, regulates certain conditions of the biosphere to keep them favourable for its continued existence is the core of Gaia Theory. This homeostatic regulatory capacity eventually weakens; it is weakening on the Earth at the moment). If the regulatory capacity collapses the conditions of the planet will radically change meaning that only very simple life-forms can survive. These simple organisms will be unable to complexify in the new planetary conditions and it will only be a matter of time before life on such a planet is annihilated. As the homeostatic regulatory ability on T weakens life starts to technologically regulate the temperature of the atmosphere in order to enable its survival. Life on T thrives and has a glorious cosmic future! What of life on NT? It gets annihilated.

I'm sure that you can see that when we leave the 'human perspective' behind and move to the 'Earth's perspective', it is clear that technological advancement to deal with the scenarios above is essential. I realise that many people who consider themselves to be 'green' or 'environmental' might need some convincing of this, particularly of the need to technologically control the temperature of the atmosphere. However, we need to be very weary of creating a dichotomy between such technological measures and either 'being green' or 'age-old wisdom'. Actions which lead to the annihilation of the life that has arisen on the Earth are neither 'green' nor 'wise'.

Objector: Thanks. I can now see the value of technology for the planet. However, our evolution in the consciousness department is lagging so badly behind our technology that we have made a total mess of the Earth, squandering its resources and jeopardising the survival of not only our own species but many thousands of others. We are so ignorant of how the planet works that we don't even know all that's in a cubic inch of topsoil. We have a lot of learning to do before we can claim that we know anything about how to take care of a planet!

NPC: I am glad that you can see the value of technology for life on the planet. I agree that we are hugely lacking in knowledge concerning the way that the planet works. However, we are learning all the time, and the way that things unfold means that we will know enough by the time that we need to know. In other words, planetary life evolved us at a stage in its unfolding when there was sufficient time for us to acquire the knowledge and technology that it requires. You seem to believe that the appropriate human actions in the realm of technology are determinable by humans and that a lack of knowledge should result in a decision not to deploy technology. But the way that things have always needed to pan out is as follows: firstly, acquire some knowledge of the world; secondly, use this knowledge to develop technologies; thirdly, become aware of the changes made by technology (the environmental crisis) and the holes in our knowledge; fourthly, deploy technology to regulate the atmospheric temperature. This is just the way that a successful life-bearing planet evolves. This process is not about pointless reckless human abuse, it is about humans being used (despite their incomplete knowledge about the planet) for the sake of life on Earth.

Objector: How can you claim that the idea that the rest of the planet's biodiversity can do very well without us is completely and utterly wrong, given that life did just fine for billions of years without us?

NPC: Consider a human who is presently not eating any food (they are on a hunger strike). I assert that: *this human cannot exist without food.* You respond: *this person hasn't eaten for 3 days. You are wrong. They can exist without food.* This isn't going to convince me. I'm still going to believe that *this human cannot exist without food.* It is a claim not about the past or the present; it is a claim about the non-immediate future. In other words, things change. You cannot make a simple extrapolation from the past to the future; this doesn't work in an evolving Solar System. You need to understand all of the relevant unfolding forces which are currently in play, the forces which are determining the present state of the planet.

Objector: The most important question is: Can we use our ability to narrate to tell better stories that won't cause our own destruction? Life will go on without us; it went on when 95% of the species were terminated and then again and again with each subsequent mass extinction. We are the ones endangered, as the dinosaurs were at one time. Everything else is just a bunch of stories to scare children around the fire.

NPC: The story that some people tell themselves that "life will go on without us" is a story which has been created on the basis of a simple extrapolation from history – it ignores the changing relationship between an ageing planet, its life-forms, and an ageing sun. It is a fiction. I am confident that in the near future this story that "life will go on without us" will be widely seen to be false.

Objector: Humans are obviously a destroying species. Thousands of species are driven to extinction through human activities each and every day.

NPC: I don't deny that human activities have resulted in the extinction of a vast number of non-human life-forms. However, it is also obvious to me that that the human species is a 'saving' species. For example, consider all of the conservation efforts which are currently going on around the world. Some human activities are 'destructive activities' and some are 'saviour activities'. The important question is: Which is the most fundamental activity and which is the most superficial activity?

In other words: Is the human species fundamentally destructive and superficially saviour? Or: Is the human species fundamentally saviour and superficially destructive? I don't deny the obvious destruction; I simply make the case that it is superficial. The human species is fundamentally a saviour species.

Objector: Geoengineering is not necessary and is too dangerous.

NPC: Many people seem to be 'scared' by the term geoengineering. In my early writings I did not use the term (I hadn't even heard of the term then!). Now I do use the term, but I only use it to refer to one particular type of activity – the technological regulation of the atmospheric temperature. Initially, this simply entails taking intentional actions to directly counterbalance past unintentional actions. A simple reversal to undo what has already been done. When you look at it this way, it doesn't seem to me to be particularly dangerous.

Let us consider the unintentional actions which need to be reversed. The human species has made massive unforeseen changes to the biogeochemical cycles of the Earth. These changes have lengthy time-lags. The effects of the changes that have been made have barely become manifest yet. For example, a massive amount of the carbon which was previously safely stored under the surface of the Earth has effectively been moved, by humans, to a temporary storage area in the oceanic thermohaline circulation. The time-lags involved mean that almost all of the carbon that has been moved in this way since the start of the Industrial Revolution is still in the thermohaline. When it re-emerges the temperature of the Earth's atmosphere is set to shoot upwards.

In short, the damage has already been done. No 'traditional' responses can provide the solution to this situation. By 'traditional' responses I mean things such as:

> Sustainable Development

> The 3 R's: Reduce, Reuse, Recycle

> Sustainable Retreat

Even a complete halt on all human activities would not provide a solution to this situation. What is the solution? The human species can take intentional actions to counterbalance the unforeseen effects of its previous activities. These activities involved releasing carbon from its underground storage areas, after which it moved to the atmosphere and the thermohaline. So, at its simplest, these intentional actions entail taking carbon dioxide out of the atmosphere. We are already making good progress at working out how to do this. This ability to offset the changes that we have unintentionally made would mean that when the carbon emerges from the thermohaline that we will be able to maintain the level of carbon dioxide in the atmosphere, and thereby prevent the temperature of the atmosphere from rising.

The future of the human species, and other complex life, depends on the geoengineering of the atmospheric temperature. So, geoengineering is necessary. Furthermore, this doesn't seem to me to be particularly scary, something we should be terrified about. Simply counterbalancing what we have already done is not so dangerous and complex as many make out. In any case, the danger has already been created; this is the solution. Compared to some of the things that we can do, such as sending technological creations to the planet Mars, such counterbalancing measures seem relatively simple. Looking a little further into the future other more-ingenious technological solutions will be required, such as blocking some of the solar input before it is able to enter the Earth's atmosphere. This is obviously a much more complex endeavour than simply removing carbon dioxide from the atmosphere.

As we imagine the Earth in the future, a planet on which life is flourishing – thousands of years into the future, hundreds of thousands of years into the future, millions of years into the future – we can see that all of the wonderfully interesting and complex life-forms that exist owe their existence to their technological saviour: the human species.

Objector: You seem to envision humans as pawns in a cosmic game.

NPC: That is a good way of putting it. We are developing the technology that life needs in order to survive because of a non-human problem, and we are currently doing it because we believe there is a solely human-created problem which needs to be solved. So, you can certainly think of humans as pawns in a cosmic game.

Objector: You say that the environmental crisis/human-induced global warming are positive events which indicate that the human species is fulfilling its role as saviour of life on Earth. This seems counter-intuitive. Clearly our actions are placing the planet in peril; how then are we its saviours? We may end up saving ourselves through geoengineering but are we saving the planet? Life on Earth will persevere irrespective of what we do. What we do seems to be largely detrimental to the living Earth. Am I missing something?

NPC: The important thing to keep in mind is that we, as a species, are always telling ourselves stories about how we relate to the rest of life on Earth, and how we relate to the rest of the universe. For example, we were created in the image of God, the Earth is the centre of the universe, we are just an advanced breed of monkeys on an average planet, we have dominion over all planetary life-forms, we are the destroyers of life and life will carry on just fine without us, we are the saviours of life (Noah's Ark), and so on. Most of these stories get rehashed and repackaged through the ages, so one is probably familiar with the typical stories even if one disagrees with them. The view of the place of humans in the cosmos that I am outlining seems to be a genuinely new view in many respects, which explains why many people find it "counter-intuitive" when they first encounter it.

Dialogue

One of the main stories that humans are telling themselves in the current epoch is: *we are the destroyers of the rest of life, but life will persevere no matter what we do. Conclusion = the rest of life would be better off if we went extinct.* It is obvious to me that this story is not true. However, in my philosophy I outline why human culture needs to go through an epoch in which the majority of people need to believe that this story is true. Such a widespread view (of human separateness, of human superiority, of humanity as the destroyer) is required for us to act in the way that fulfils our purpose as the saviours of life on Earth.

Without us the life that has arisen on the Earth, particularly the complex life, doesn't have much of a future; this is just a straightforward reality of existing in an evolving solar system/cosmos. Most people can easily understand that this is the case in the long-term. I am explaining why it is also the case in the immediate future. In short, we are the bringers forth of the technology which saves, the bringers forth of the Earth's technological armour. In order to save we need to impart a certain amount of destruction; we need to destroy in order to save.

Objector: Can you expand on the idea that humanity is separated from the rest of life/the cosmos due to the environmental crisis?

NPC: Sure. We are currently living through an epoch which can be described in various ways: the environmental crisis, the era of human separateness, the era of technological birthing. The bringers forth of technology on a planet (the human species) are separated from their surroundings – separated from the rest of planetary life and separated from the rest of the universe – through their technological creations. In turn, these technological creations are the cause of the environmental crisis. So, human separation is intimately connected to the environmental crisis.

The epoch of human separation extends back to the time when advanced tools/technology were first brought forth, whereas the epoch of the environmental crisis only extends back to the time when technology became so pervasive as to cause worrying large-scale effects. However, as the planet entered the epoch of the environmental crisis, the degree of perceived human separateness reached a new level of extremeness (due to the increased pervasiveness of technology). The human species *is* the bringer forth of technology, and technology brings forth an environmental crisis. So, to be the human species *is* to bring about an environmental crisis.

The human separation entails an automatically generated 'sense of specialness' (a sense that the human species is superior) within humans. And, this 'sense of specialness', in turn, leads to a conception of the non-human as 'other'/'nature'. When we have fulfilled our purpose and entered the era of sustainability, then the era of separateness will come to an end. Planetary life divided itself for its own benefit; the separation was needed to bring forth technological armour for the benefit of life as a whole.

Objector: I am not sure I understand your claim that the human species has a 'purpose' as 'saviour of life on Earth'. I take it you mean that there is a metaphysical purpose, but this requires an 'end', in the

Aristotelian teleological sense – an end, moreover, that has somehow inhered in us from our origins. The only grounds on which I can imagine someone accepting this idea is God, e.g., God created us for a purpose and now we are on the brink of realizing that purpose.

NPC: You are right that I believe that the human species has a metaphysical purpose, a purpose that is independent of what humans tell themselves about whether or not their species has a purpose. However, my view has no need for God; all it requires is that there is directionality in the cosmos from non-life to life, from life to human, from human to technological globalised society/environmental crisis, and from environmental crisis to active technological regulation of the temperature of the planetary atmosphere. You say a metaphysical purpose requires an 'end'. In my philosophy the end is U*. U* is the state in which the entire universe is pervaded with (human) life. It should be stressed that having an end does not entail that the end is reached! Whilst having no need for God, my view is perfectly compatible with the existence of an intelligent creator of the universe.

Also, I should stress that even if one believes that the universe is wholly devoid of purpose, one can still believe in the main conclusions/outcomes of my philosophy. One can still believe that it is in the interests of planetary life for humans to deploy technology to actively regulate the atmospheric temperature. The existence of the two forms of global warming, the weakening of the planetary homeostatic regulatory capacity, the time-lag effects which are a force for immense future global warming, all of these things create the reality of the situation we find ourselves in. All of these things exist even if the universe is wholly devoid of purpose. So, our choice of path (Path 1, Path 2/3) needn't have anything to do with the notion of purpose. However, if one sees the purpose in the universe then this lends considerable support and beauty to the need for Path 2, and ultimately Path 3.

Objector: What do you think of environmental activists? Are they wasting their time?

NPC: I think environmental activists are doing a great thing if they are doing what they are truly passionate about. I don't think they are wasting their time; they are part of the *force to environmental sustainability,* which is a currently relatively weak but important force. However, it should be realised that there are different types of environmental problems. It should also be realised that there is a crucial distinction between aspiration and reality. When it comes to the specific environmental problem that is human-induced global warming, then the overwhelming majority of environmental activists are supporters of *Path 1*. This means that I believe that their global warming activism is grounded in flawed assumptions. In other words, if their dreams came to pass – the dreams which underpin their activism – the result would be tragic, the devastation of the wonderful life that has arisen on the Earth. Environmental activists typically have a philosophical worldview which sees the human species as the destroyer of planetary life and they therefore seek to move to a future in which human 'interference' with 'nature' is minimised. So, whilst I am glad that there are environmental activists in the global warming domain, I am also glad that they are sufficiently ineffective that that their aspirations will not become reality. In the global warming domain we need Path 2/3 not Path 1. In the other, non-global warming environmental domains, I am a great supporter of environmental activism. In these domains there is typically no gap between the aspirations of environmental activists and what would be good for life on Earth.

Objector: Why do you think that the vast majority of environmental activists have a flawed view when it comes to global warming?

NPC: This isn't just an issue relating to environmental activists; the vast majority of humans have a flawed view when it comes to global warming. This is because the global warming issue has, so far, been dominated by the science of global warming; there has been no real engagement with the philosophy of global warming. Environmental activists typically seem to be humans who are particularly close to that which is largely 'untainted' by humanity (trees, oceans, countryside, mountains, non-human animals, wilderness, plants) and this closeness generates an emotional response, a desire to protect the non-human from the human 'destroyers'. I believe that this emotional response can be transcended, and when this happens that one can come to appreciate what is really in the interests of life on Earth. If this occurs then one will move from a flawed view to an unflawed view (one will see why Path 2/3 is in the interests of life on Earth).

Objector: You propose that individual humans contribute to the evolutionary progression of human culture/life on Earth through their individual actions. I don't see how my life contributes in this way; I am just an ordinary person.

NPC: Not every human contributes to the evolutionary progression of life on Earth/human culture, but the overwhelming majority do. If you drive a car, if you live in a modern house, if you buy food at the supermarket, if you travel by train, or airplane, or bicycle, or helicopter, or boat, or bus, then you are part of the fundamental force which is propelling life on Earth towards its glorious destination. This force is simply the process of feeling states yearning for, and moving to, a higher state of 'goodness'. I sometimes refer to this force as having two elements – *the force to environmental destruction,* and *the force to environmental sustainability.* Activities such as those that I just mentioned are part of *the force to environmental destruction* element.

If you withdrew from society completely and lived by yourself in a forest where you foraged for all of your food/supplies/needs, then you would not be contributing to the evolutionary progression of human culture/life on Earth. If you lived this way for a sufficiently long period of time, and had a non-technological way of life, you might even cease to be a part of the human species and would return to being an individual non-human organism. I should stress that this applies to an individual, or individuals, that withdraw from society completely. If such an activity led to wider cultural changes, such as the setting up of an eco-community and educational centre which humans from afar could visit, then the human/s living in the forest would be part of *the force to environmental sustainability.*

So, it is exceptionally likely that, through your individual actions, you are contributing to the evolutionary progression of human culture/life on Earth.

Objector: The planet and its life-forms, far from ageing, are in their prime. In a few billion years, when the ageing of the planet and the ageing of the Sun becomes a problem, we will have run our course and, based on history, better, smarter, and more adapted species than ours will already have taken the helm.

NPC: You are expressing a widely held but false view, the belief that the ageing of the *Earth-Sun-Solar System* is only of relevance for life on Earth in "a few billion years". The reality is that it is of relevance now. The planetary homeostatic regulatory capacity has already severely weakened (due to the ageing of the *Earth-Sun-Solar System*). What this means is that if, in the near future, there is a large jump in the GMST into a range in which complex life on Earth is unable to survive (due to non-human causes or non-human causes exacerbated by human causes) then all planetary life is doomed. In other words, if the human species becomes extinct then life on Earth will not be able to recover; it will be fatally wounded. The simple life-forms that remain would not be able to complexify, life would not be thriving on the planet anymore; planetary life would be clinging on to survival whilst awaiting its inevitable extinction. The human species (the bringer forth of technology) will be at the helm of any 'mature' planet where life is thriving.

Objector: You do not seem to take seriously the manifest risks of planetary geoengineering. Personally, I am terrified of the prospect. We don't have much of a track record of positively transforming our environment – quite the contrary. Geoengineering may be the final act of hubris that destroys us.

NPC: I understand that there are risks involved; however, the opponents of geoengineering typically believe that the risks are of a much greater magnitude than they actually are. Furthermore, the alternative to taking the risk is certain extinction for life on Earth. So, the options are to take a risk or to die; when one can see this then one will surely embrace the risk. You say that you are "terrified" of the prospect. I am not sure if you really understand what is being proposed. This is an area in which some humans have set out to terrify other humans, so it is perhaps not surprising that you are terrified. We've massively disrupted the carbon/biogeochemical cycles already; this is what you should really be terrified about. The homeostatic regulatory capacity of the Earth (its ability to support complex life) has been seriously weakened due to non-human causes; this is what you should really be terrified about. Using technology to reverse what we have already done (simply pulling some carbon out of the atmosphere), and also to provide the solution to what we haven't done, is a good thing; it isn't something to be terrified about, it is something to be celebrated. However, there are those who are 'anti-technology' who have an agenda of appealing to deep-seated fears that using technology to regulate the atmospheric temperature would lead to catastrophe and should be avoided at all costs. People are often driven and deeply affected by these deep-seated fears and emotions. Appreciating our place on the planet, our purpose, our specialness, can help these fears and emotions to be overcome.

Objector: If it wasn't for a random event the dinosaurs could still be ruling the planet; it is only because of this random event that humans now rule the planet.

NPC: This is obviously a very contentious, although seemingly widely-held, view. The general point is this: Is there directionality in the evolutionary process towards the 'human'? And, a secondary question of importance is thus: What is the 'human'? I take the 'human' to be 'the bringer forth of technology' and believe that on every successful life-bearing planet the 'human' evolves. Your comment on the dinosaurs is not persuasive enough to convince me that this is not so. I accept that there are random events which affect life on Earth. However, the trunk of the evolutionary progression of life on Earth is not fundamentally knocked off course by random events.

Objector: I don't believe that there is purpose in the universe; so, the human species cannot have a purpose. There is therefore no need for geoengineering.

NPC: Even if there is no purpose in the universe, if the human species evolved as some kind of fluke and there is no underlying direction to the evolution of life on Earth, then there is still a need for geoengineering. The important point to accept is that there is a need for geoengineering for the sake of the future survival of both the human species and the life that has arisen on the Earth. When this is accepted then there is scope for discussion as to whether or not the universe is purposeful, and, if so, whether the purpose of the human species is to geoengineer the temperature of the atmosphere.

Objector: You keep going on about purpose and now you seem happy to admit that the universe might not be purposeful.

NPC: I will clarify the relationship between geoengineering and purpose. My main claim is that the human species needs to use technology to actively regulate the temperature of the atmosphere and that this is a positive wonderful event for life/the Earth (it isn't a 'weapon of last resort' which is how such a measure is standardly seen). This main claim is embedded in a philosophical worldview in which the universe is purposeful. However, the main claim is itself independent of the issue of purposiveness. In other words, accepting that the universe is not purposeful wouldn't fatally undermine the main claim. So, the universe being purposeful can be thought of as a second claim which bolsters the geoengineering claim without potentially fatally undermining it. I am convinced that the universe is purposeful and this is why I keep referring to purpose; however, if one is not convinced by this one can still accept the truth of the main claim.

Objector: You seem to be an advocate of the overman/superman, and believe that humans should just continue with their pursuits with no fear or concern for the consequences.

NPC: I believe that there are 'two' major forces running through human culture. The first is characterised by domination and control (no fear for the consequences), the second is concerned with sustainability (concern for the consequences). I call these forces *the force to environmental destruction* and *the force to environmental sustainability*. Both forces are good and valuable, and it is the way that they interact that achieves the optimum outcome. Firstly, *the force to environmental destruction* dominates and brings forth the birth of the technology that life requires. Secondly, *the force to environmental sustainability* grows in strength and brings forth a sustainable human culture. This interaction and transition occurs automatically through the vast majority of humans living largely in accordance with their inner feeling states (which are determined by the evolving Solar-Systic structure, see *Chapter Ten* and *Chapter Eleven*).

Objector: Don't you think that given the critical thinking faculty it is mandatory for humans to carefully consider the potential pitfalls of their actions and to consequently choose appropriate methods for survival and living?

NPC: I agree. Indeed, the purpose of this book can be thought to be to enable people to engage their critical thinking faculty in the domain of global warming, rather than to have opinions which are based on emotional responses and very limited information. The choice of "appropriate methods for survival" that you refer to is the choice of path that I outline in this book – Path 1 or Path 2/3.

Objector: Not only would geoengineering strategies be very costly, they could have many foreseen and unforeseen harms, without being much help in reducing global warming. The solution might turn out to be worse than the problem. To me it doesn't make sense continuing our profligate, wasteful lifestyle and spending even more money on things that might not work, when we could just change our lifestyle, turn off a few lights not in use, etc., and reduce several environmental and other harms while saving money. It's a total no-brainer.

NPC: Your reasoning concerning the options we face makes sense given the starting assumptions that your process of rationalisation flows from. The problem is that your starting assumptions are too few; in other words, if you considered a wider range of relevant facts then you would reach a different conclusion. Firstly, the need for geoengineering primarily stems from non-human induced global warming, so addressing our "profligate wasteful lifestyle" is not a replacement for geoengineering in addressing this need. Secondly, the biogeochemical cycles of the Earth are currently in a state which involves very large yet to be manifested changes waiting to be unleashed resulting from past human actions; addressing our "profligate wasteful lifestyle" obviously cannot deal with these changes. In other words, what seems to be obviously the correct option to you – avoid geoengineering and reduce our profligate wasteful lifestyle – is actually a path to disaster.

Objector: I am a big fan of Professor Franz de Waal. He claims that anything humans can do that we cling to as evidence of 'human exceptionalism' is done by numerous other mammals. We just aggregate it at a bit more efficient level. In the end, from his perspective, there's nothing that sets us apart and we have no 'special' role to play.

NPC: We (Franz and I) are, I believe, in total agreement about 'human exceptionalism'. I don't know if he has read my first book *(Is the Human Species Special?)* but I am sure he would totally agree with everything that I say in *Chapter Two*. Franz is primarily concerned with the relationship between humans and non-human primates; we *both* claim that there is nothing 'special' about humans when you consider this relationship in isolation from a philosophical worldview. However, there is a second issue which exists at a deeper level. I have been dealing with issues concerning the nature of the universe and the way it evolves, this is where the 'specialness' of the human species comes in for me. This specialness does not result from a comparison of the attributes of a human with the attributes of a non-human (which is all Franz is up to); it results from a philosophical worldview of the universe and the way that the Solar System evolves through time. If Franz considered these issues, and my arguments that the human species is the zenith of the evolutionary progression of life on Earth, then I think there is a good chance that he would share my conclusions.

So, I think you are wrong to conflate two things; firstly, the fact that Franz and I believe that there is nothing that sets humans apart when we directly compare humans to non-human Earthly life-forms; secondly, that humans have no special role to play on the planet. The first does not necessarily entail the second. The first is a direct comparison between two life-forms which operates without any philosophical baggage; it is not embedded within a philosophical worldview. The second is a philosophical view of the evolving Earth/Solar System and a claim that part of evolving 'planetary life' is of particularly immense significance in that unfolding process.

Objector: 97% of the species which have ever lived on the Earth have already disappeared. Humans will go extinct soon too.

NPC: I don't think that humans should be compared to the species which have previously gone extinct. Consider the possibility that the human species is not just another species – it is the zenith of the evolutionary progression of life on Earth, the bringer forth of technology, the moulder of its surroundings *par excellence.* The 97% figure isn't going to convince me that the human species is going to go extinct in the same way as non-technological species have. Species often go extinct because their environment changes/habitat disappears; technology can stop the environment changing and provide replacement habitat. Species often go extinct because of a lack of mobility; technology provides super-mobility. Species often go extinct through predation; the technological species is the top of the predation pyramid.

Objector: But I have watched the film *The 11th Hour.* The wise men and women at the end, especially the elder from the Onondaga Iroquois, makes the point that life on Earth will continue, with or without humans, and that there is plenty of time for the planet; however, we are running out of time to save a version of advanced civilization. 97% of all species have gone extinct, and there are plenty of reasons to think humans will, too!

NPC: I haven't seen the film, so I don't know if these humans you refer to proclaim themselves to be "wise", but from the words you have ascribed to them they seem to be far from wise. It is surely true that without humanity life on Earth will continue for the foreseeable future in the form of microbes; however, without humanity planetary life as a flourishing entity containing a plethora of interesting complex life-forms will be doomed. There isn't "plenty of time for the planet" if we are talking about a planet flourishing with life. Such a view (expressed by the "wise men and women") arises because of a lack of appreciation of the bigger picture, particularly of such things as planetary astrobiology and the currently severely weakening homeostatic regulatory capacity of life on Earth.

Objector: All other species share habitat. Humanity is the only one that deliberately destroys habitat so that almost nothing else can live there. That is obviously evil. What we are doing to the Earth proves that we are not a moral species, nor intelligent.

NPC: We can ask ourselves whether taking the habitats of other species and driving other species to extinction is a bad thing. Taken in isolation this seems to be a bad thing. However, if this transformation of habitats is part of the evolutionary trajectory of planetary life as it moves from simple beginnings (when its future existence was precarious) to technological manipulation of its surroundings (to ensure its future existence), then such transformations are clearly in the interests of planetary life. This means that humanity is not obviously evil; it is, in fact, the saviour of planetary life. So, what we are doing to the Earth can certainly be seen to be a "moral" doing carried out by an "intelligent" species.

Objector: History and the present is showing us ample evidence that we are damaging the environment thanks to our cosy life. If humans were the ultimate species, the end of any evolutionary chain and a sure bet on any thriving planet, then we should show responsibility, care and respect for the surroundings and all other beings. Do we?

NPC: You are missing the point. You are right that I believe that the human species (the bringer forth of technology) is the zenith of the evolutionary progression of a thriving life-bearing planet. However, what this entails is actually *a lack* of care and respect for our surroundings (the opposite of what you assume); a lack that is caused by a lack of awareness. In other words, the age of technological development is characterised by a separation between the human and their surroundings, a separation which means that humans are busy doing their thing – investigating, moulding, shaping, transforming and exploring their surroundings – in the absence of the awareness that they are harming their surroundings, and in the absence of the awareness that they are acting in an unsustainable manner. The awareness of the unsustainability, and the awareness of the separation itself, the awareness of the lack of awareness, follows the age of technological development. In other words, in order to be the zenith of the evolutionary progression of life on a planet the human species needs to have a lack of care and respect for its surroundings and for other beings.

Now, of course, human culture goes through different epochs and the era of technological development which is characterised by a lack of care and respect, will be followed by another era. I believe this era will be characterised by care and respect for both the planetary biosphere as a whole and for all individual planetary life-forms. You pose the question "Do we?" and the answer is both yes and no. In other words, some humans do and some humans don't. However, when we look at the collective actions of the entire human species, then during the era of technological development (which is still ongoing) then the answer is 'no'. When we move to the coming era the answer will be 'yes'.

Objector: When I look at the planet I see 7 billion hairless apes running amok.

NPC: I am not sure what you are seeing! The cultural trajectory from the origin of the human to globalised technological society was an inevitable trajectory. Individual humans are pawns/puppets/cogs in this cultural-evolutionary progression; they are not, in any meaningful sense, running amok.

Objector: You claim that the human species is the zenith of the evolutionary progression of life on Earth. Why should I believe this? In the future a species might evolve which is much more intelligent than the human species. This species would then surely be the zenith.

NPC: The zenith is not determined by intelligence. The zenith is marked by life on Earth reaching the stage at which it becomes technological. The zenith obviously entails a certain level of intelligence; indeed, the bringing forth of technology requires a high level of intelligence. It is surely very interesting that research indicates that human brains have evolved to what is known as the 'evolutionary sweet spot' – the perfect balance between high intelligence and a balanced personality. This research implies that life-forms cannot sustainably become more intelligent than humans currently are. The universe has seemingly provided humans with just enough intelligence to fulfil their purpose. This research implies that the technological zenith is simultaneously the intellectual/intelligence zenith. This seems to be more than just a happy coincidence. However, if a species of life were to evolve on the Earth (in the past, present or future) which is more intelligent than the human species then this wouldn't change the fact that the human species is the zenith.

Objector: You seem to believe that we can justifiably continue to damage our surroundings due to a blind belief that whatever be the new situation, we and our technology will always eventually find a way, sooner or later, to deal with the situation. I believe this to be an extremely dangerous way of thinking.

NPC: My main objective is to simply give a philosophical account of the ways things are, the way things have been, and the way things will be in the future. The human species is that part of life on Earth that has become technological. Technology is required to save life on Earth from extinction. Technology also causes problems which have non-technological solutions. This is just the way things are. I don't have a 'blanket' view concerning technology. I am not a personal fan of technology (I have never been in an airplane; I don't like cars; I cannot stand microwaved food). I don't believe that technology provides the solution to all problems. I have simply been provided with the insight that the short-term survival of complex life on the

Earth requires technology in one particular domain; technology is required to actively regulate the atmospheric temperature.

Objector: We are utterly embedded in the natural world and we are dependent on nature; in contrast, we are not dependent on technology, or economics, or science.

NPC: Nature is an 'opposition term' which has been created by humans who have become separated from the fundamental oneness of the cosmos. There is no human "dependency on nature"; humans *are* nature. Technology is not 'non-natural'; economics is not 'non-natural'. Those who see the utter embeddedness of humans in the cosmos have no need for such oppositions, no need for the terms 'non-natural' and 'natural'.

Objector: You claim that the purpose of the universe is to evolve life and that the purpose of life is to stay in existence. But to discuss purpose on the subject of the universe is really to set your own rules and then try to form what reality you perceive so it fits into your procrustean bed of what life should look like. This is a variation on creationism and intelligent design, with really no basis in logical thought – it's more like religion.

NPC: I think you're missing the point. This seems to be a very logical and rational thing to believe, and has nothing to do with religion, or creationism, or intelligent design. One could believe that all life wants to die (the opposite of my belief that life wants to stay in existence). But this belief has no basis in rational thought. Indeed, there are many reasons to believe that it is false, so only an 'illogical' human would believe such a thing. The body seeks to repair itself when damaged. We continuously strive to protect life and extend life (cryogenic freezing!). Does one feel happy when lots of life-forms die in a natural disaster? Surely one doesn't. The continued existence of life is a good thing – a good state for the universe. To deny this seems to be illogical. So, I'll continue to believe that life wants to stay in existence, that its purpose is to survive. I will also continue to believe that life is a precious and rare part of the universe and that the bringing forth of life is the purpose of the universe.

Objector: I cannot retain your optimism that the up and coming *force to environmental sustainability* will help to ensure future survival despite the increasing population.

NPC: If you are not optimistic that sometime in the future humans will be living sustainably (as *the force to environmental sustainability* becomes stronger) then you believe that humans will never be able to live sustainably and that the human species will therefore go extinct. I hope that you don't really believe this. *The force to environmental sustainability* only evolved in the mid-1900s, yet it already has global reach and, despite its weakness, it is continuously growing in strength.

Objector: Humans have never lived sustainably in the past, so there is little likelihood that they will in the future.

NPC: Your view is flawed. It is only within the past century that humans have realised that there is an 'environmental crisis', that there is a need to live sustainably at the planetary level. All of history up to this realisation has been characterised by a lack of effort to live sustainably. You cannot reasonably extrapolate from a long period of 'no effort to live sustainably' and then claim that humans will not be able to live sustainably in the future! Achieving sustainability first requires realisation of the need, then it requires changes to make this a reality. Cultural structures move slowly so you cannot expect an instant move to complete sustainability following the realisation of the need.

Objector: You claim that the human species is the "saviour of life"; does this imply that all life would eventually cease to exist were it not for some actions by the species *Homo sapiens,* i.e., that no life would ultimately exist except for certain proactive efforts on the part of a single, late appearing species?

NPC: Yes, that is right. All of the life that has arisen on the Earth will *eventually* cease to exist if the human species does not ensure its continuation through the deployment of technology.

Objector: The Earth will go on without us for billions of years, so our conception of mankind being important for the Earth is, to say the least, arrogant.

NPC: The Earth will go on for billions of years without us, but this isn't the point. The point is whether the Earth will be able to sustain complex life in the near future without us. It is obvious to me that it won't. If the human species were to go extinct tomorrow then this would herald the end for complex life on Earth in the near future, and ultimately the end of all life on Earth.

Objector: Surely some life will survive. What do you mean by "complex life"?

NPC: Complex life-forms are all of the 'large' life-forms which would disappear from the Earth if the GMST increased significantly; think of 'animals' and 'plants'. These life-forms need a particular temperature range in order to survive. However, if the GMST becomes too high for complex life there will still be life on the planet in the form of microbes. Life can usually find a way to survive in some form even if the conditions which enable it to thrive on a planet have forever been lost.

Objector: So, you accept that life is incredibly difficult to eradicate once it has appeared. If the human species becomes extinct microbes and cockroaches will persist with no trouble at all. It is then possible that a second technological civilization may evolve from some other lineage.

NPC: You are right that life is incredibly difficult to eradicate; this is why I make a distinction between life on Earth and complex life. However, I am convinced that the Earth/Solar System is already at the stage in its evolution which means that if the human species and all other complex life on Earth becomes extinct that there simply won't be enough time for a second technological civilization to evolve from some other lineage. This means that the extinction of the human species is ultimately the extinction of (all) life on Earth. Furthermore, if the extinction of complex life is caused by a large jump in the GMST, then this means that the new planetary conditions won't be suitable for *any* complexification of life (given the new relationship, the new balance of forces, between the *Sun-Solar System-Earth*).

Objector: The ecological crisis will bring about ever greater suffering for an ever greater number of people. Our death as a species will surely take place during the current century and it will be extremely painful.

NPC: Why are you so pessimistic? We can probably all agree that the 'ecological crisis' has resulted in some suffering, and will result in some suffering in the future. However, to paint a picture of ever increasing suffering leading to the death of the human species within the next 86 years seems a tad dramatic. The ecological crisis as a concept only arose around 60 years ago, and in that very short space of time we have already made massive progress at working out how to reverse the changes that we have unwittingly made. It is quite amazing really. In the near future, probably within 100 years of the concept arising, we will be able to pull carbon out of the atmosphere, (re)store it underground, and thereby actively technologically control the temperature of the atmosphere. There is no need for ever increasing suffering and death.

Objector: I totally disagree that we can continue to behave as we have been behaving in the last century and yet survive as a species.

NPC: I am inclined to agree with you (so you are not disagreeing with me). My view is that we have behaved as we have (in an unsustainable way) so that we will fulfil our purpose, which is to technologically regulate the temperature of the GMST. This will only become crystal clear and widely accepted in the future when humans look back into our current epoch – the epoch of unsustainability/technological development – and see that it was a necessary epoch in order to fulfil the purpose of the human species. Human actions on the planet go through three stages: 1) sustainable; 2) unsustainable; 3) sustainable. This is the way that a successful life-bearing planet evolves.

Objector: If we believe that reduced consumption, alternative energy, and all the other good green things won't happen then we are in trouble. If we actually believe that they won't happen, then we have an excuse for doing nothing. Doomers are inevitably doomed. Therefore, hope and optimism, the Polyanna approach, is the only approach that has any chance of solving the problem of global warming. We are sure to perish if we don't believe.

NPC: There is some truth in the idea that something is unlikely to be achieved if it is believed that it cannot be achieved. However, one needs to have hope in the right thing; one needs to have the right objective. It is of little use having hope and optimism in a false idol, in a road to nowhere. There are two paths facing humanity and you assume that the first path is the correct path; in reality, this is the path to destruction. If, despite reality, and despite what is good (technological regulation of the temperature of the Earth's atmosphere) you won't let go of your belief that everything will be fine if consumption is reduced and alternative energy sources escalate, then you have an erroneous and dangerous view. A view that is optimistic and hopeful for sure, but a view that is also mistaken and dangerous.

Objector: I think techno-quick-fixes should be put into a similar category as the second coming of Christ, the arrival of the angels or aliens to save us, or any of those scenarios that allow humankind to dodge their responsibility in favour of handing it over to a 'higher power'.

NPC: Your opinion is wrong. Technologically regulating the temperature of the atmosphere isn't a "quick-fix" it is a long-term necessity. Achieving this objective is the reason the human species came into existence and it is our responsibility to achieve it for the benefit of planetary life. *Not* achieving this objective would be dodging our responsibility.

Objector: I believe that humankind's influence on Earth's life is pretty mild; species come and go just like they always did.

NPC: The thing to be concerned about is the speed with which we seem to be changing habitats and changing the planet, compared to the extinctions of the past. Some of humankind's influence on Earth's life is obviously apparent (cities, motorways, agricultural land, etc.). However, there are yet-to-be-manifested time-lag effects resulting from humankind's activities which seem set to cause much more significant changes in the coming centuries than these obviously apparent changes. Compared to the pre-human changes that occurred on the Earth, there are two unique aspects of humankind's influence. Firstly, we now *know* the impacts of our activities. Secondly, we are that part of life on Earth which has become technological. This means that humankind's influence on the rest of life on Earth needs to be seen in a fundamentally different light to pre-human changes to life on Earth.

Objector: As we only know a little bit about the planet we live on, and a whole lot less about other potentially life supporting planets in the universe, how do you substantiate the claim that "technology is the sign of a healthy thriving planet?" Humans may turn out be a 'failed' evolutionary experiment.

NPC: We know enough to know that without technology life will emerge on a planet and then become exterminated: emergence, extermination; emergence, extermination. Technology can break this cycle: emergence, survival. After a particular stage in Solar-Systic evolution, technology enables life to survive and thrive on its original planet; it also ultimately enables life to survive and thrive in other places. As survival is a better, more optimal, state than extermination, technology is a sign of a healthy thriving planet.

Objector: Humankind has language and symbolic thinking, this makes us special.

NPC: I really don't see how language/symbolic thinking could, in itself, meaningfully be said to make the human species special. All species have unique attributes, but it is far from obvious to me that language/symbolic thinking is a unique human attribute, let alone a 'special-making' attribute. Furthermore, you seem to be advancing an anthropocentric attitude in believing that human attributes are special whereas non-human attributes are not special. If you believe that language and symbolic thinking are uniquely human, and that they make humans special, then you need a convincing argument as to why this is so. It is no good just asserting the statement.

Objector: As a species as a whole, we are not able to sing in tune, so the chance that we will disappear fairly quickly seems to me more feasible than the possibility that we will successfully technologically regulate the atmospheric temperature.

NPC: The beauty of the cosmos is that desired outcomes can be attained even though the agents who bring about the outcomes do not consciously set out to achieve these outcomes. Furthermore, it is never the case that advances in human culture/society are brought about by the entire species "singing in tune"; advances are brought about by a small subset of the whole. So, it doesn't matter if the human species as a whole is unable to sing in tune, the technological regulation of the atmospheric temperature can still be successfully achieved.

Objector: Geoengineering! Can we try it on some other planets first to see if it works? The trouble with geoengineering is that we have no ways of predicting its effects or its side-effects.

NPC: Our inability for perfect prediction is certainly a problem. However, there is barely anything that we can perfectly predict. We generally predict things with certain degrees of confidence and our predictive ability typically improves with experience. The real "trouble" is that without geoengineering all complex life on Earth dies in the near future (and ultimately all planetary life dies).

Geoengineering is not necessarily something which is particularly dangerous or unpredictable. When certain clouds naturally form they reduce the GMST by raising the planetary albedo. If humans geoengineer the atmosphere by putting 'sun shades' in the atmosphere then they are simply replicating cloud activity. I don't know of anyone who thinks that our inability to perfectly predict when clouds will form is something that we should be particularly worried about. So, you shouldn't worry about geoengineering because of the human inability to perfectly predict things.

Furthermore, the initial geoengineering activities that are required are simply the pulling of carbon dioxide out of the atmosphere. It is the high levels, and the increasing levels, of carbon dioxide in the atmosphere which results in unpredictable side-effects. So, geoengineering, the pulling of carbon back out of the atmosphere, is a stabilisation measure. It is simply a reversing of what has already been done. It is a 'cancelling out' of that which will otherwise cause unpredictable and undesirable side-effects. In other words, far from being the cause of unpredictable side-effects, geoengineering is fundamentally the preventer of unpredictable side-effects.

Objector: Many people are critical of geoengineering because it provides a techno-fix, a way of putting off doing the most intelligent thing, which is to start changing the way that we think and the way that we act, so that we can emit less fossil fuels.

NPC: This is the way that the issue is typically framed; geoengineering is seen as a 'weapon of last resort' which we might need to turn to if we are unsuccessful at doing "the most intelligent thing", which is acting in a way that involves massively reduced fossil fuel emissions. It is important to realise that my approach and conclusions come from a completely different perspective that does not sit easily with this framing – a broader cosmic evolutionary approach relating to the nature of the Solar System/Universe and the way that it evolves.

From my perspective, human geoengineering of the atmospheric temperature is a cosmic inevitability and a positive event in the evolution of both life on Earth and the Solar System. Whilst there are uncertainties concerning the achievement of this end, I believe that humans are resourceful enough to be able to achieve it. Let us consider the state of planetary evolution a few hundred years ago. If a human then asserted that *it is inevitable than humans will walk on the moon*, it is easy to imagine other humans thinking that such a feat is fantastically improbable, full of uncertainties and dangers. They respond: *How could we possibly do that?* However, this feat was inevitable because humans are that part of the evolving planet that has become technological. This feat was also successfully achieved despite the uncertainties and dangers. Technological advance is a long process which entails exploration at many levels (including of the moon) and ultimately leads to the regulation of the atmospheric temperature.

So, your claim that "the most intelligent thing" we could do is to "change the way we think and act" misses the mark. We are already doing the most intelligent thing. We are bringing forth the technology that life needs in order to survive and thrive; then, when this is done, we will fully transition to a sustainable way of living.

Objector: You say very little about how the technological regulation of the temperature of the atmosphere is to be achieved. This is surely a crucial omission.

NPC: There are 2 issues:

1) Philosophical/Scientific = Is there a need for geoengineering? Is geoengineering a desirable outcome which should be actively striven for?

2) Engineering/Scientific = How do we implement geoengineering?

Dialogue

There are experts in universities and other organisations across the world working on 2). However, at the moment environmental groups and some academics are campaigning against geoengineering research and experiments. The result of this is that governments have (temporarily) prohibited some of the required research.

My concentration is on 1) above. The reason for this is that I seem to have some fairly unique things to say about 1), and it is 1) which is seemingly the most important of the two issues. This is because if enough people could be convinced that there is a need for geoengineering then things would be much easier regarding the research and resources for 2). The second issue is simply an engineering issue; whereas, 1) contains a plethora of deep and intricate issues — philosophical, ethical, moral, and scientific. I have got enough on my plate with 1). So, I feel that I can leave the simpler 2) to the experts who are working on this.

Objector: Your views are dangerous because the possibility of climate engineering can blind people to our moral responsibilities.

NPC: I don't know where your idea of "our moral responsibilities" comes from. It is obvious to me that our moral responsibility is to save life on Earth from extinction, and this requires utilising technology to regulate the temperature of the Earth's atmosphere.

Objector: Isn't your view just the same as the Dark Mountain view.

NPC: I came across the Dark Mountain view very recently. There seems to be some overlap between this view and mine. The similarity lies in the belief that no actions that humans can plausibly take (in the absence of geoengineering) can prevent an environmental collapse. According to the Dark Mountain view the environmental movement is disingenuous when it suggests that with concerted effort (through Path 1) we can avoid an environmental collapse; the conclusion is thus reached that we should focus on dealing with the global disruptions that will result from not addressing contemporary environmental issues. This is thus quite a broad view *which is focused on environmental adaptation.*

I reach a very different conclusion and have a more narrow focus. I agree that the environmental movement is wrong in its assumption that an environmental collapse can be averted if we tread Path 1. However, I hope that the environmental movement, in the global warming domain, will come to realise the error of its ways. Furthermore, and in any case, humanity will prevent an environmental collapse through technologically regulating the temperature of the atmosphere. There are other environmental problems, and resource shortages are likely in the future, but I don't think that there will be an environmental collapse in the future. This is very different to the (Dark Mountain) view that a collapse is inevitable and we should thus focus on preparing for it. My focus is not on adaptation, it is on using technology to prevent an environmental collapse from happening in the first place.

Objector: Do you think that we have a responsibility to 'future generations'?

NPC: I prefer to think of the 'future existence of planetary life', rather than of 'future generations'. However, I will answer your question in terms of 'future generations'. The answer is: yes. If we ignore 'future generations' of humans and non-human life-forms, if we focus solely on the interests of the life-forms that are currently alive on the Earth, then we can consider what the most appropriate course of action would be. Given that there are some uncertainties and risks in the realm of geoengineering, and given that the serious time-lag effects resulting from previous actions don't start until much later this century, there would be a strong case for not geoengineering the temperature of the atmosphere. In other words, if we are wholly selfish, and are only concerned about our own existence (the existence of life-forms currently in existence), then the optimal course of action seems to be to avoid all possible risks to our own existence.

However, as soon as we bring 'future generations' into consideration then everything changes. If we care about the future existence of life on Earth then we need to take responsibility for the future. We need to accept the risks and uncertainties arising from geoengineering in order to enable the existence of 'future generations'. In short, we owe it to 'future generations' to geoengineer the temperature of the atmosphere. To not do so would be extremely selfish.

Objector: Is your view scientific or religious?

NPC: The core of my philosophy is grounded in science, yet it didn't arise from science. My philosophy is fully compatible with science and it is embellished through science. I think the dichotomy between science and religion is unhelpful. It seems to me to be more useful to contrast scientific knowledge with non-scientific knowledge. Science cannot truly be divorced from non-science. All science is ultimately grounded in particular assumptions; these might be philosophical, religious, spiritual, or metaphysical. Furthermore, these assumptions are often unrecognised. It is because of the influence of these non-scientific assumptions that scientific data is often interpreted very differently. So, my philosophy inevitably entails both the scientific and the non-scientific.

My philosophy has very little to do with organised religion, although I believe that there are some truths in ancient religious texts that reflect the nature of the relationship between the human species and the non-human life-forms of the Earth. For example, the notion of human dominion is common in ancient religious texts, and the idea of humanity as the technological saviours of non-human planetary life-forms is one way of interpreting the biblical account of Noah's Ark. I am also open to the possibility that the entire universe is the creation of an intelligent entity, the qualities of which we cannot know (one might, or might not, want to refer to this entity as God).

Dialogue

Objector: According to Heidegger our fantasies of control and our alienation from the rest of nature is what got us into the environmental mess in the first place.

NPC: We are alienated from the rest of nature (it is this alienation that *creates* the concept of 'nature'), but we are not in a mess. We are bringing forth technology for the benefit of life on Earth.

Objector: If it is our 'nature' to attempt to control 'nature', then it is equally our and its 'nature' to fail.

NPC: Your starting assumption seems to be that minimal human interference with the non-human is desirable for life on Earth. This assumption is wrong and it explains why you mistakenly see human control as inevitably leading to failure.

Objector: We must move away from the anthropocentric worldview that our forefathers perpetuated; a worldview which led to the abuse and destruction of so many aspects of our biosphere.

NPC: What you say isn't in conflict with my view. I agree. I simply seek to convince you that "the anthropocentric worldview of our forefathers" has a function, a positive role to play in the evolution of life on Earth. You seem to believe that this worldview is a wholly negative thing, rather than a positive necessity on a thriving planet.

Objector: Almost all of the non-human species that currently live on the Earth were living on the planet for millions of years before humans evolved; this gives them priority over us. To allow human numbers to rise to levels that threaten other species is criminal.

NPC: If A exists before B this does not entail that A has priority over B. One needs reasons why the prior existence entails priority. I have presented lots of reasons to believe that the prior existence of non-human life-forms *does not* give them priority over the human species.

Objector: I'm a fan of living nature-neutral: Don't take more from nature than you give back. So, if a certain gas comes from taking substances out of the Earth and burning them (we free that gas, so to speak), then we should take the same amount of that gas and bind it, for instance.

NPC: Your view is grounded in the assumption that we are a 'guest' on the planet and that we should seek to perturb our host as little as possible by having a "neutral" impact on the planet. Throughout this book I have been making the case that this 'guest' assumption is wrong. Far from being a guest we are the zenith of the evolutionary progression of planetary life. Our purpose, our task, is *not* to be nature-natural. Our purpose is to master 'nature' to the extent that we are technologically able to regulate the temperature of the Earth's atmosphere for the benefit of planetary life. However, I am glad that you can see the sense in geoengineering, glad that you can see that it is sensible rather than scary.

Objector: According to numerous religious texts (such as the Bible: Genesis 1:26) all of humankind is given permission to have dominion, take charge of, rule over, be head over, reign over, have complete authority over, have power over, be responsible for, or be sovereign to, all the other animals on the Earth. However, given our present knowledge of environmental issues we should surely reject this view.

NPC: No. We shouldn't reject this view. This view is correct. What needs to be rejected is the way that the environmental crisis has come to be widely conceptualised; that is, as entailing that the human species is the destroyer of life on Earth. The reality is that the human species is the saviour of life on Earth and this is very closely intertwined with having dominion over all non-human planetary life.

Objector: Being the saviour entails having dominion?

NPC: Exactly. The bringing forth of technology entails dominion.

Objector: I am glad to learn from you that human-induced global warming is not such a big deal after all because it'll have the positive outcome of controlling the temperature of the Earth's atmosphere through geoengineering. However, the human species is not damaging biodiversity in only one way, i.e. through global warming. So, how are we going to save our planet from all of the other environmental threats that we are inflicting biodiversity with?

NPC: Attaining a sustainable and harmonious future for life on Earth requires the interaction of various factors. Emerging technologies will play a role, new behaviour patterns will play a role, and a crucial factor will be a widespread reconceptualisation and realisation of the impacts of our actions. I believe that 'spiritual growth' will play a large role in achieving a sustainable and harmonious future. Finally, the way that we see the human/non-human relationship, the way that we see our place in the cosmos, is important; in this respect, I hope that this book might help to bring about a more sustainable and harmonious future.

Objector: You claim that humans suffer because they became separated from 'nature' and thereby became destined to endure suffering, as if no other sorts of beings ever had or ever would suffer; I have to say, that's only in your imagination.

NPC: You are severely misunderstanding my view. I believe that the human species suffers more than any of the other species of life-forms on the Earth, but this is not to say that no non-human Earthly life-forms suffer. It is surely exceedingly obvious that a multitude of non-human life-forms suffer! I simply claim that the human species suffers *more* than any other part of life on Earth.

Objector: The lesson we need to learn urgently is this: we cannot do without the rest of the planet's biodiversity, but it can do very well without us.

NPC: The major theme running throughout this book is that this assertion is completely and utterly wrong. Non-human planetary life cannot do very well without us. I suggest you go to the start of the book and start reading it.

Objector: I think we should use resources on poverty alleviation rather than geoengineering.

NPC: There is no need to set up a dichotomy here. An immense amount of resources are currently being used *trying to avoid* geoengineering. These resources could be more fruitfully used on both poverty alleviation and geoengineering. There are enough resources for both poverty alleviation and technologically regulating the temperature of the Earth's atmosphere.

Objector: If human-induced global warming is not actually occurring, or if it is occurring but is of trifling insignificance, would that undermine your view?

NPC: No, it wouldn't. I believe that a significant part of the human species needs to believe that human-induced global warming is a very significant threat to the continuation of the human species. However, whether it is actually such an immense threat is not so important. The belief is more important than the reality. Either a (false or correct) belief in the reality and seriousness of human-induced global-warming, or the actuality of (human-induced or non-human-induced) global warming, could be the stimulus which causes the human species to fulfil its purpose of technologically regulating the temperature of the Earth's atmosphere (to combat non-human-induced global warming).

So, if we were to technologically regulate the temperature of the Earth's atmosphere and then attain scientific proof that human-induced global warming never existed, that would be perfectly compatible with my view. After all, the human purpose is to address the problem arising from non-human-induced global warming.

Furthermore, if there was a widespread realisation within the human species concerning the nature of its purpose, then no actions or beliefs in relation to human-induced global warming would be required. We could just get on with fulfilling our purpose in the full awareness of its existence and of what we were doing. However, the way that the planet unfolds through time seems to entail the human species acting so as to fulfil its purpose and then at a later time, when the purpose has been fulfilled, realising that it had a purpose. This means that the actions which lead to the purpose being fulfilled will be grounded in false beliefs, such as: the belief that the human species does not have a purpose, the belief that the human species is the destroyer of life rather than its saviour, and possibly the belief that human-induced global warming is a threat to the continuation of civilization (this might or might not be a false belief).

If tomorrow it was scientifically proven that human-induced global warming doesn't exist, *and* this resulted in the human species not technologically regulating the temperature of the Earth's atmosphere, then this would undermine my view. However, *anything* that leads to the human species technologically regulating the temperature of the Earth's atmosphere supports my view.

And, of course, human-induced global warming is actually occurring, and it is a phenomenon of immense significance. However, we should keep in mind that the main threat to life on Earth is non-human-induced global warming.

Objector: I presume then that if it was proven that *non*-human-induced global warming does not exist, that your view would be completely and utterly wrong.

NPC: Yes. You are right. At the very heart of my philosophy is the conviction that non-human-induced global warming exists. It is the existence of non-human-induced global warming which engenders the human species with its special place as the zenith of the evolutionary progression of life on Earth. If there was no non-human-induced global warming then the human species would be just one life-form among many.

Objector: It is conceivable that humanity decides not to geoengineer the temperature of the Earth's atmosphere. If this is so, then your philosophical worldview is wrong.

NPC: This might be conceivable. However, I don't believe that in reality we will able to not geoengineer the temperature of the Earth's atmosphere. It is possible that humanity decides not to carry out such an activity, but then our hand will be forced to do it anyway. When it is widely realised that our very survival depends on such an activity then we will do it; at this stage previous decisions will be irrelevant.

To appreciate why the technological regulation of the atmospheric temperature is inevitable it is useful to envision the human species as a collection of puppets which ultimately acts in the interests of life on Earth (and the Solar System, and the Universe) whether they realise it or not. It isn't possible that we could have created a world substantially different from the world we live in today (globalised, capitalistic, technologic); we simply don't have that much freedom. In other words, the technological regulation of the temperature of the Earth's atmosphere is going to happen whether we like it or not. The only choice we have is to consciously embrace this outcome and see it as a joyous event for planetary life, rather than to resist it. If we achieve this feat then we can make the best out of the situation and minimise the suffering experienced by both human and non-human life-forms.

Objector: Can you accept that your philosophical worldview could be wrong?

NPC: Sure. On a rational level I can, at times, doubt almost anything – the existence of an external world, purpose in the universe, directionality in evolution, the correctness of almost all my beliefs. However, on a non-rational level, the sense of inner conviction that I have that my view is obviously correct is immense, beyond doubt, beyond question; there is no doubt that it is right.

PART 3: ARTICLES

Was the Cosmic Bringing Forth of Humans 'Inevitable'?

I would like to briefly consider the question of whether the cosmic bringing forth of humans was inevitable. This question, when fully elucidated, clearly has an answer (either yes or no); however, it is questionable whether humans can know with certainty what the answer is. To make progress in attempting to answer this question we first need to elucidate what it means. There are two elements to the question. Firstly, what is a 'human'? Secondly, what does it mean to talk of 'inevitability'?

Let me start by considering the notion of 'inevitability'. According to contemporary conventional wisdom the cosmic bringing forth of humans was not inevitable because the paths which biological evolution takes are not 'directed' towards a particular outcome. According to this view, biological evolution is simply a process through which the 'fittest' life-forms survive. If one believes this then it is hard to also believe that one particular life-form – the 'human' – was 'inevitable' from the first moment that life evolved on the Earth. One can respectably believe that biological evolution is driven by the 'survival of the fittest' and also that there is a tendency for biological evolutionary paths to head towards complexity. However, to believe this is to believe a very different thing from believing that the heading towards complexity is a heading towards one particular form of complexity – the 'human'.

So, according to contemporary conventional wisdom the 'human' is not inevitable because the evolutionary paths that life has taken on the Earth could have been very different. You will probably have heard people assert something along the following lines:

> *Things could have turned out differently. If the asteroid that wiped out the dinosaurs had missed the Earth then the evolution of life on Earth would have been very different; dinosaurs could still be the dominant life-form on the planet and humans would never have evolved.*

There is still debate about what exactly caused the end of the 'dinosaur era', but the general point which is being made is simply that biological evolution is a process which is pervaded with contingency. According to this view, life is evolving in a particular direction and then a random event such as an asteroid strike radically changes that direction. This means that it makes no sense to say that from the moment that life first emerged on the Earth the evolution of the 'human' was inevitable.

There is clearly a sense, due to random events, in which the past (and present and future) evolutionary paths which life has taken are not inevitable. The biological evolutionary paths could surely have been different. However, accepting that these paths could have been different is compatible with the view that they are 'inevitable'. This is because to believe that biological evolutionary paths are 'inevitable' is simply to believe that they are heading in a particular direction to a particular goal. Random events can be seen to be outside influences which can temporarily cause a deviation from the pre-existing evolutionary trajectory. When they are temporarily knocked off course the evolutionary paths then reassemble and head back towards the direction which they were previously heading in. So, to believe in 'inevitability' is not to believe that evolutionary paths could not have been different, it is simply to believe that evolutionary paths are heading in a particular direction to a particular goal. In other words, even if we accept the idea that a random event such as an asteroid strike wiped out the dinosaurs, we can still assert that it was inevitable that the 'human' would evolve.

Consider an analogy. John is attempting to drive from Glasgow to London, and he is absolutely desperate to get to London (let us imagine that his life depends on it). When he leaves Glasgow there is a sense in which it is 'inevitable' that he will arrive in London. This 'inevitability' does not entail that the path which John takes to get from Glasgow to London is inevitable. If things go smoothly then he will take the route he planned in advance. But he could encounter roadblocks and/or accidents ('random events') which cause his path to be very different to that which was planned in advance. Despite his desperation it is also not inevitable that he will arrive in London (he could die in an accident on the way).

So, to believe that the cosmic bringing forth of the 'human' was 'inevitable' is to believe that the biological evolutionary paths of life on Earth were always heading towards the 'human'. There are many possible paths to this destination, just as there are many routes from Glasgow to London. It is entirely sensible to say that it is 'inevitable' that John will make it to London, despite the minute possibility that he will die in an accident on the way. Similarly, it is also correct to say that it is 'inevitable' that the 'human' would evolve on the Earth, despite the minute possibility that a random event such as a massive asteroid strike could have wiped out all life on Earth before the 'human' had a chance to evolve.

Let us move on from 'inevitability' and consider that which is hypothesised to be 'inevitable' – the 'human'. What is a 'human'? The answer *seems* to be obvious: humans are a species of animal which inhabit the Earth; they typically have two arms, two legs, a torso and a head. This is how we typically think of the 'human' as a member of a particular biological species – the 'human species'. I don't have this conception of 'human' in mind – simply a member of a biological species with a head, torso, arms and legs – when I consider whether the 'human' was 'inevitable'. I have in mind a different conception of 'human': the essence of what we call 'humans' is not their torso, arms, head and legs. The 'human' is marked out by the way that it sees itself compared to its surroundings, it is marked out by the particular way that it thinks, it is marked out by its actions – such as engaging in science and bringing forth technology.

When I say that the 'human' was 'inevitable' I don't mean that a particular biological arrangement of limbs was 'inevitable'. I mean that this way of seeing itself, this way of thinking, this bringing forth of technology, was inevitable. Was the cosmic bringing forth of the human 'inevitable'? Yes.

Two Routes to the Need for Geoengineering

My belief that the evolution of the 'human' was 'inevitable' is intimately connected to my belief that humans need to geoengineer the temperature of the Earth's atmosphere. However, there are two different routes which one can travel which transport one to this destination. I would like to briefly introduce you to these two routes.

The first route is embedded within a particular view of the universe as a whole. One can believe that the entire universe is a purposefully evolving and unfolding entity and that geoengineering is a part of this unfolding. According to this view, the universe seeks to evolve life wherever possible and life then sets out on its own journey towards complexification, maintaining itself, technology, geoengineering and ultimately spreading out from its host planet to other parts of the universe. This view entails that the evolution of the 'human' was 'inevitable'. You will find an example of the view that the universe is a purposefully evolving whole when you reach the final article: *Friedrich Hölderlin and the Environmental Crisis*.

The second route is free of any philosophical baggage concerning views of the universe as a whole. This route is wholly science-based; it does not entail any views concerning 'inevitability, 'purpose', or the nature of the 'human'. It is simply a scientific fact that the weakening planetary homeostatic regulatory capacity, in tandem with the increasing output of the Sun, necessitates the need for geoengineering of the Earth's atmospheric temperature if life is to survive and thrive on the Earth in the future. Interfusing with this need are the indisputable perturbations which humans have made to the biogeochemical cycles of the Earth, but which have not yet become manifest in terms of the warming of the atmospheric temperature (the elastic band has become nearly fully stretched but has not yet been released). These interfusing perturbations further necessitate the need for geoengineering. If this doesn't occur then all complex life on Earth seems certain to die when the effects become fully manifest in the near future (seemingly well before the year 3000). The current limits of human understanding concerning the way that the biogeochemical cycles of the Earth operate mean that other interpretations are possible. One could believe that we have thousands and thousands of years to geoengineer the temperature of the Earth's atmosphere, or one could believe that the need is in the next few decades.

I will be further considering both of these routes in the following articles. I should stress here that the second route whilst self-standing, is also a part of the first route. In other words, the human scientific and technological project, the associated environmental crisis, and the human realisation of the scientific need for the geoengineering of the temperature of the Earth's atmosphere, are important stages in the evolution of a purposefully unfolding Earth/Solar System/Universe.

The Need for Geoengineering

I would like to say a little more about the need for geoengineering. If the life which has arisen on the Earth is to survive there is no doubt about this need. The only real question which is of concern is the timescale of the need. To appreciate this we need a wide perspective on the 'environmental crisis' and the human presence on the Earth.

It is easy to make the mistake of believing that before humans had a significant impact on the Earth that the Earth was in some kind of 'static' state. According to this view, the Earth (and life on Earth in particular) was in a perfectly 'stable' state until humans started having a significant impact on the Earth; in other words, humans are taken to have 'disturbed' a pre-existing static/stable state and thereby destabilised the systems and cycles of the biosphere.

Whilst there is a grain of truth in this view – the human impact on the Earth has been significant compared to the impact of the non-human life-forms that came before – it is essential to appreciate that the idea that the pre-human Earth was in some kind of 'static' state is a myth. The Earth is an evolving and ageing whole. When the Earth was in its 'youthful' stage of evolution life spread out over the entire planet and was almost effortlessly able to maintain the conditions of the Earth to keep them favourable for its continued existence. The Earth aged out of its youthful stage a long time ago. Life on Earth is now in a stage (at an age) in which it is in peril (see Sir James Lovelock on this – *Gaia, The Ages of Gaia, The Revenge of Gaia*). In the distant future the ageing process will have reached the stage in which the Earth is inevitably unable to sustain any life. The youthful exuberance of a young Earth gives way to decay and death. There are thus three stages to the evolution of life on a planet:

1 Youthful Exuberance

2 Perilous Existence

3 Inevitability of Death

The 'human' (the technological animal), as a highly complex arrangement of life, inevitably arises in *Stage 2: Perilous Existence.* Our purpose, our unique ability, is to face the peril head-on. The pre-human non-living forces are driving the planet back to a lifeless state. The human has the ability, on behalf of life, to extend *Stage 2* for an immense period of time. This is wondrous as a state of life is immensely better than a state of death.

I have been referring to the 'inevitability' of death. You might be wondering whether there is a way in which the Earth, and its life, can survive forever. This is a possibility, but it requires the human species to be able to technologically move the Earth as a whole from our solar system to a more youthful part of the universe. This is not impossible but it is also a highly fanciful possibility, and in its absence the Earth will inevitably disintegrate when the Sun expires. There is a less fanciful possibility; it is much more plausible that life on Earth will survive by utilising human technology to leave the Earth and the Solar System. The creation of a highly advanced spaceship to transport life in this way bears a striking resemblance to the biblical account of Noah's Ark. The survival of the life that has arisen on the Earth in the distant future clearly depends on human technology.

Let us now consider the more immediate survival prospects for the life that has arisen on the Earth, given that we are currently living through *Stage 2: Perilous Existence.* This is a complex issue with many aspects; however, there is one thing that is crystal clear. If there is no geoengineering of the atmospheric temperature then the Earth will become devoid of life much sooner (and seemingly immensely sooner) than if such geoengineering occurs. The non-technological ability of the ageing Earth to keep the conditions of the Earth favourable for the flourishing of life is already severely weakening due to the ageing process. However, just as with an individual human, technology can extend life. If humans can successfully geoengineer the Earth's atmospheric temperature then the life-forms of the Earth can look forward to a long and rosy future. If not, there is just a downward spiral of decay and death. In the absence of such geoengineering we hastily move into *Stage 3: Inevitability of Death.*

There is no doubt about the need for geoengineering; the only question is when the activity is required. If the human perturbations to the biogeochemical cycles of the Earth hadn't occurred then there would probably be quite a few thousand years before complex life on Earth died. However, the human movement of massive amounts of fossil fuels from their underground storage areas to the temporary storage areas of the slow moving deep ocean currents (the thermohaline circulation) leads me to believe that we have, at most, until the year 3000 to be successfully geoengineering the temperature of the Earth's atmosphere. However, I could be being very optimistic here. It is best to fully acknowledge the uncertainties relating to the timing of the need for technological control of the atmospheric temperature. It could be that the future survival of complex life on Earth requires such technological control much sooner than this, possibly even before the year 2100.

There is no certainty about the date by which geoengineering is required; there is just certainty about the need.

The Nature of the Universe

In order to fully appreciate why there is a need for geoengineering, and why the purpose of the human species is to fulfil this need, it is helpful to think about the nature of the universe and the place of life within the universe.

What is the nature of the universe? You might believe that humans are simply not able to meaningfully answer this question. Indeed, it would seemingly be foolish to claim that one knows with one hundred per cent certainty what the nature of the universe is. Nevertheless, humans clearly have access to some aspects of the universe; after all, they are part of the universe, and they are also able to observe certain movements within the universe. On the basis of such access and observation it is reasonable to draw some conclusions concerning the likely nature of the universe.

One of the main conclusions that one can draw is that the universe is 'purposeful'. There is much to be said about what exactly this means, and why your own experiences relating to both what goes on inside your own body, and what you observe in the world, suggests that this is so. In short, to believe that the universe is 'purposeful' is to believe that the universe has a particular aim/objective/goal (this doesn't entail that the universe is minded, or conscious, or aware, or thinking). Furthermore, to believe that the universe is 'purposeful', that it has a particular aim/objective/goal, is simply to agree with Aristotle that:

> There is something divine, good, and desirable... [that matter] desire[s] and yearn[s] for
>
> (cited in Skrbina, 2005, *Panpsychism in the West.* London: The MIT Press. p. 46)

So, what is the nature of the universe? The universe is an entity which "desires and yearns" for something "divine, good, and desirable". It is this desiring/yearning which underpins the evolution of the universe (in both the 'living' and the 'non-living'). Therefore, the purpose of the universe is to attain a 'good and desirable state', to attain that which is 'yearned for'.

What is it that is yearned for? What is this good and desirable state? What is yearned for is life, the bringing forth of life and the continuation of life. This continuation requires the bringing forth of the human (the technological animal). So, the human is that part of life, that part of the universe, which is yearned for the most. The human is the state of ultimate goodness, the most desirable state for the universe to be in.

Links between My Philosophy & the Buddhist Theory of Atoms

I would like to say a little more concerning the nature of the universe. Recently I was scanning through some books and I discovered a passage which is in very close accordance with one of the more audacious claims that I make in my book *An Evolutionary Perspective on the Relationship Between Humans and Their Surroundings* (2012).

In this book I consider how the qualities of humans compare to the qualities of everything in the universe (to put it rather crudely). By 'everything in the universe' I mean both the basic constituents of the universe (we can refer to these basic constituents through labels such as 'atoms' or 'ultimates') and everything that these constituents form when they come into particular arrangements – things we call 'stars', 'tables', 'oranges', 'seagulls', 'submarines' and 'humans'.

Humans are clearly part of 'everything in the universe' but do they have qualities which 'everything in the universe' does not? To be clear, to talk of 'everything in the universe' having a particular quality is to talk of *every* thing in the universe having this quality; no thing would lack this quality. If humans have quality B*, but 'everything in the universe' also has quality B*, then humans are clearly not distinguished from 'everything in the universe' through B*.

From the evolutionary perspective, in the realm of perception, I make the case that humans only have two senses which distinguish them from 'everything in the universe'. Compare a human with an atom and consider the five traditional human senses. What distinguishes human from atom? What distinguishes the human is that it has two senses (seeing and hearing). The other three human senses (touch, taste, and smell) are also qualities of the atom. So, in the realm of perception, humans are distinguished from 'everything in the universe' because of the possession of two senses. This view is part of my overarching philosophy.

Let us now consider the passage which I discovered in a book by the Dalai Lama (*'The Universe in a Single Atom'*, 2006, Abacus, p. 55):

> The early Buddhist theory of atoms, which has not undergone major revision, proposes that matter is constituted by a collection of eight so-called atomic substances: earth, water, fire and air... and form, smell, taste and tactility... an 'atom' is seen as a composite of these eight substances, and on the basis of the aggregation of such composite 'atoms', the existence of the objects in the macroscopic world is explained.

So, according to the Buddhist theory of atoms, it is the case that 'everything in the universe' is smelling/touching/tasting. Therefore, this theory is in accordance with my philosophy. In terms of the senses/perception, humans are distinguished from 'everything in the universe' through the possession of two senses. *Every*thing in the universe has the other 'three' senses (in *An Evolutionary Perspective* I make the case that these three traditional senses are actually a singular sense). This is a hard thing for many to make sense of. Indeed, in the above work the Dalai Lama claims:

> Personally, I have never understood the idea that qualities like smell, taste and tactility are basic constituents of material objects. (p. 58)

and, therefore that:

> this aspect of Buddhist thought... must now be modified in light of modern physics' detailed and experimentally verified understanding of the basic constituents of matter in terms of particles such as electrons revolving around a nucleus of protons and neutrons. (pp. 58-9)

I am greatly saddened that the Dalai Lama should personally feel the need to abandon the long-standing Buddhist theory of atoms on the basis of modern physics. For, modern physics has nothing whatsoever to say on the issue of whether or not the Buddhist theory of atoms is true. Modern physics has in no way proven that 'everything in the universe' does not smell/touch/taste. Indeed, as I urge, the *Evolutionary Perspective* (which is partially grounded in modern science) gives us good reason to believe that the Buddhist theory of atoms has been right all along.

The claim that 'everything in the universe' has the quality of smell/taste/tactility is possibly a claim that you find hard to take seriously. After all, as we have seen, even the Dalai Lama cannot understand it. Nevertheless, it is obvious to me that 'everything in the universe' has the quality of smell/taste/tactility, and I find comfort in the fact that, long ago, the formulators of the Buddhist theory of atoms were also able to see that this is so.

The GreenSpirit Journal Comments on ITHSS

You might be interested in some comments which were made concerning my first book: *Is the Human Species Special?: Why human-induced global warming could be in the interests of life* (2010). The comments were made in the *GreenSpirit Journal* (2011, 13:3) in an article entitled *'A New Fire, a New Mind'* which was written by June Raymond. Here is what she had to say:

> Another thing that left me thinking was a book I read recently which asks some very challenging questions about what we as a species are doing here on this planet. It deals with some very deep and radical questions about our role in creation and in the future of the Earth. The proposal that the author makes is that we were put here so that when the Earth becomes too hot, because of the sun's heating up, we will be able to save our planet through our technology; for example we might be able to create satellites with mirrors which could reflect the sun's heat away... And so to return to the question, 'Does the human species have a purpose? and if so what is it? Is it to rescue life on planet Earth and if so how? The conclusion, that Earth created us to rescue it when the sun becomes too hot, and the present global warming is going to help us get our act together in preparation for this, is in terms of the best modern scientific thinking, not unreasonable.

The GreenSpirit Journal Comments on ITHSS

I consider the view that I initially outlined in ITHSS — *that the environmental crisis and human-induced global warming are positive events in the evolving cosmos; that the human species has a purpose: to technologically regulate the Earth's atmospheric temperature* — to be quite radical. Up until now, these events, the environmental crisis and human-induced global warming, have been universally portrayed as negative events. So, I am happy that I have seemingly been able to present the view in a way that makes it seem "not unreasonable" and "compatible with modern scientific thinking". This was my primary objective in writing ITHSS, to introduce the view and to make it plausible ("not unreasonable"). The next stage is, of course, to persuade people who are already open to the plausibility of the view, that the view is actually *the most plausible view*. This is my task in this book.

The First Book Critiquing ITHSS

In the previous article I shared with you an excerpt from the first journal article (that I am aware of) that refers to my work. In this article I am going to share with you an excerpt from the first full-length book that responds to my work. This book was written by Peter Xavier Price from the Sussex Centre for Intellectual History and is entitled: *'Human Specialness': The Historical Dimension & the Historicisation of Humanity* (2012). Price provides an interesting critique of my first book: *Is the Human Species Special?: Why human-induced global warming could be in the interests of life* (2010). Here is some of what he has to say:

> What is it about humanity that places it far above other life-forms? Why does it often perceive itself to be so unique when the natural world is teeming with biological anomalies? Perhaps even more tentatively, can humans truly claim to be the remedial agents destined to solve the current global environmental crisis? In Neil Paul Cummins' recent book, Is the Human Species Special?, the author sets out to address these very questions by speculating that mankind is indeed special because it represents the pinnacle of the evolutionary process. Employing a radical thesis which bears a remarkable resemblance to the infamously distorted dictum of the Vietnam War (i.e., that of 'destroying the village in order to save it'), Cummins suggests that mankind has reached a paradoxical stage in its development, whereby its imminent

downfall may suddenly prove to be the means of its ultimate redemption. Thus, in this swashbuckling interpretation of the human response to environmental uncertainty, Cummins paints a picture of the human condition as seemingly analogous to the closing act in a grand, teleological narrative of biological endeavour and primordial purpose. 'Could it be', he speculates, 'that in order to fulfil its purpose and be the saviour of planetary life ... humanity had to believe that it was potentially the destroyer of planetary life?'

From the outset, it is important to note that Cummins' publication is an accomplished work – at once entertaining as it is erudite. The author clearly exhibits the full depth and range of his innate interdisciplinarity as he weaves seemingly disparate strands from his economic, environmental and philosophical background into a tightly argued and well-constructed piece. But what, we may be entitled to ask, are the inherent pitfalls to the bold thesis that he has constructed? Indeed, some may even believe that it falls short at the first hurdle. For how, they might argue, can the wiping out of a whole village constitute any sort of liberation for its inhabitants? Yet, as valid as this criticism may appear to be on the surface, it should be acknowledged that Cummins does in fact cover his tracks in this respect when he proposes that it is the imminence of the environmental disaster (rather than the purported disaster itself) that will ultimately ensure the planet's survival. Therefore, as far-fetched as the overarching argument may appear to be to some, it is simply wrong to accuse the author of outright contradiction.

This essay, then, is in large part an attempt to sketch out a far more convincing alternative to Cummins' arguments; but not, as may be expected, to what is essentially the central argument contained therein. In doing so, it aims to redeploy Cummins' ideas and to use them as a catalyst for further discussion; though, perhaps, in a direction that he mostly neglects or even ignores. At this initial stage, and in the interests of brevity, we may wish to describe this endeavour 'an assessment of the relative absence of history in Cummins' idiosyncratic account of human specialness'. For, appositely, this essay also seeks to highlight the importance of recognising humanity's unique sense of its own historicity – and, by extension, the decisive role that this must surely play in any adjudication of what it is to be an exceptional species. It is hoped, therefore, that we have already gone some way towards accounting for the choice phrases (i.e. 'historical dimension' and 'historicisation of humanity') which both comprise the frontispiece to this work. Nonetheless, what they mean in precise terms should become increasingly transparent as the essay develops. Suffice it to say that, having achieved this, we will then be in a much better position to review the suppositions undergirding Cummins' work.

(pp. 5-9)

Indeed, Cummins' shortcomings are even further compounded by his exploitation of a number of schemes within his thesis which, as we have shown, are demonstrably historical, and yet do not appear to be historically accounted for. For it surely cannot have escaped notice that Mandeville's early account of wealth-creation, via the paradox of 'unsocial sociability', bears more than a passing resemblance to the author's bio-evolutionary (or even quasi-eschatological) account of the potentially redemptive qualities of 'fallen' man. A similar case may even be inferred by his adoption of decidedly Malthusian concepts, about which, again, there appears to be no acknowledgment at all. Yet, even more significantly, Cummins' account of what he calls the 'trajectory of human evolution from hunter-gatherer to technological society' — indeed, the very thread upon which his whole argument is based — appears, in truth, to be little more than the eighteenth-century Scottish 'four-stages-theory', albeit in slightly modified form. Had Cummins acknowledged this interesting fact, he might even have reached the conclusion that we may now be entering (or already find ourselves in) a quinquennial, climatical phase of a potential 'five-stage theory', replete with its own conundrums and challenges. Since he does not, it is with deep regret that the author seems so unable to construct a thesis containing greater reference to, and perhaps greater reverence for, crucial historical antecedents. For, if he had done so, it certainly would have been that much more difficult to dispute so many of the arguments contained therein.

(pp. 40-2)

I am grateful to Price for his consideration of how my philosophy fits into the broader development of the history of ideas. As I explained in the early part of this book, my philosophy is largely 'self-generated' rather than an extension of pre-existing ideas. I am unaware of large parts of the history of ideas and the heart of my philosophy did not grow out of an awareness of the previous ideas of others. This is why large parts of my philosophy are not "historically accounted for". If people with a greater knowledge than me of the history of ideas are able to fit my philosophy into a broader historical context, a context in which human understanding concerning our place in the cosmos gradually evolves, then that seems to be a very valuable endeavour.

In particular, I like Price's suggestion that my philosophy extends the Scottish 'four-stages-theory', and that we could now be in a: "quinquennial, climatical phase of a potential 'five-stage theory', replete with its own conundrums and challenges."

Ahead of the Curve

The debate between those who believe that geoengineering the temperature of the atmosphere of the Earth is inevitable and those who believe that it can be avoided seems set to intensify in the near future. One characteristic of this debate is that it is standardly carried out within narrow conceptual parameters. In contrast to this narrowness, my philosophy is very broad in scope; for example, it involves the various stages of the evolution of the Earth and the Solar System. From this wider perspective, the current epoch of technological development reveals itself as a particular stage in the evolution of a successful life-inhabiting planet. When one appreciates that the bringing forth of technology is in the interests of the totality that is life on Earth, then one will clearly see that this is of great significance to the geoengineering debate. In other words, my philosophy provides a unique perspective on the geoengineering debate through expanding the scope of the debate.

It is clear to me that life on Earth is currently in a state of great excitement as it brings forth the technological armour which will help to ensure its future existence. A large part of this technological armour is the active technological regulation of the atmospheric temperature. This is of great importance to the geoengineering debate because it means that geoengineering should not be seen, as it traditionally is, as a 'weapon of last resort', as something which is a bad thing, as something which only has to be resorted to because of the damage that humans have created due to their 'selfish ways'. In our bigger picture we can see the cosmic inevitability of the outcome. We can see that this inevitable outcome is a good thing which is in the interests of life. This is our unique perspective.

The geoengineering debate is typically framed in terms of how human societies, and their environmental impacts, could change in the immediate future. On the anti-geoengineering side of the debate is the belief that: *we can change, we can reduce our future impacts to such an extent that geoengineering can be avoided.* Those who are pro-geoengineering typically reply that: *this is exceedingly unlikely, if not impossible, in the required timescales for change.* The debate is thus framed around how much human societies can change their future environmental impacts. I have tried to convince you that geoengineering the temperature of the atmosphere is the goal which life has striven to attain for millions and millions of years. From this perspective it is inevitable. The only question is when people *en masse* will come to realise that this is so. My unique perspective on the debate seems to be 'ahead of the curve'. However, I am glad to see a gradually increasing realisation that the geoengineering of the atmospheric temperature is inevitable. I hope that soon there will also be a growing realisation that such an outcome is not something to be undertaken with regret, but that it is an outcome which is a cause of celebration for all life on Earth.

Here are some examples of the growing realisation of the inevitability:

1. In the Tulsa Law Review (Volume 46, Spring 2011, No. 2) Jay Michaelson presents a paper which is entitled: *Geoengineering and Climate Management: From Marginality to Inevitability.*

2. In May 2012 there was a conference at the University of Oxford (Institute for Science and Ethics) entitled: *'Geoengineering: Science, politics and ethics'.* This is the conference introduction:

With the failure of international negotiations, global greenhouse gas emissions are now on a trajectory that is worse than the worst-case scenario. As a result, climate scientists are beginning to contemplate a response to climate change that has previously been taboo, geoengineering — the intentional, enduring, large-scale manipulation of the Earth's climate system. This series of six lectures will cover the broad range of issues raised by the emergence of climate engineering as a response to climate change.

You will see from the conference introduction, as is always the case, that the geoengineering "result" is seen to be a response to "failure". Instead, let us see the result as a cause for celebration, as the outcome which life has been seeking to attain for a very long time!

The Need for a New View of Humans in the Cosmos

After I wrote *Ahead of the Curve* an article appeared in *The New Yorker* which addresses the question of whether there is a plausible technological fix to global warming. This article reinforces the point that I was making; it also contains some material which concerns me. Here is an excerpt:

> Until recently, climate scientists believed that a six-degree rise, the effects of which would be an undeniable disaster, was unlikely. But new data have changed the minds of many. Late last year, Fatih Birol, the chief economist for the International Energy Agency, said that current levels of consumption "put the world perfectly on track for a six-degree Celsius rise in temperature. . . . Everybody, even schoolchildren, knows this will have catastrophic implications for all of us."
>
> (The New Yorker, *'The Climate Fixers: Is there a technological solution to global warming'*, Michael Specter, 14 May 2012)

The growing realisation of what we have done, what we need to do in the immediate future to rectify this (technologically regulate the temperature of the Earth's atmosphere), and that this is a good thing for life on Earth because it is a solution to non-human-induced global warming, is the core of the position that I have been outlining since my first book *Is the Human Species Special?: Why human-induced global warming could be in the interests of life* was published in 2010.

This excerpt from *The New Yorker* is evidence that we are now progressing through the stage of 'growing realisation of what we have done'. This stage will be followed by the second stage: 'growing acceptance of the need for the active technological regulation of the temperature of the Earth's atmosphere'. The third stage: 'the widespread realisation that this is a good thing for life on Earth', seems to be still a long way off. Indeed, in *The New Yorker* article, Professor Hugh Hunt (Trinity College, Cambridge) who is working on geoengineering solutions for regulating the temperature of the atmosphere states:

> I don't know how many times I have said this, but the last thing I would ever want is for the project I have been working on to be implemented... If we have to use these tools, it means something on this planet has gone seriously wrong.

I understand why people have this view, the view that if we technologically regulate the Earth's atmospheric temperature something "has gone seriously wrong". They have this view because they are 'behind the curve'. This view pervades contemporary thought and is unquestioningly assimilated. But this view is wrong. It is unhelpful. It is dangerous. If we don't replace it with a *New View of Humans in the Cosmos* there will be harmful consequences for life on Earth, and for the human species.

It is our destiny, our purpose, the very reason we came into existence, to deploy/implement the technologies which Professor Hunt is working on. The deployment of such technologies would mean that everything on this planet *has gone seriously right*. The time of implementation would be a time of great ecstasy and excitement for life on Earth! Yet those who are developing the solutions which life so badly needs do so with a sense that what they are doing is a 'last resort', something done out of desperation, something done with a sense of regret! How nice it would be if these people, and the wider public, could appreciate that they are the saviours of life on Earth. They should be treasured and protected. They should be extremely proud of what they are doing.

Those that are 'way behind the curve' are a danger to the survival of life on Earth; such dangerous views are revealed in *The New Yorker* article:

> Last fall, the SPICE team decided to conduct a brief and uncontroversial pilot study. At least they thought it would be uncontroversial. To demonstrate how they would disperse the sulfur dioxide, they had planned to float a balloon over Norfolk, at an altitude of a kilometre, and send a hundred and fifty litres of water into the air through a hose. After the date and time of the test was announced, in the middle of September, more than fifty organizations signed a petition objecting to the experiment, in part because they fear that even to consider engineering the climate would provide politicians with an excuse for avoiding tough decisions on reducing greenhouse-gas emissions. Opponents of the water test pointed out the many uncertainties in the research (which is precisely why the team wanted to do the experiment). The British government decided to put it off for at least six months.

> "The scientist's focus on tinkering with our entire planetary system is not a dynamic new technological and scientific frontier, but an expression of political despair," Doug Parr, the chief scientist at Greenpeace UK, has written.

> "When people say we shouldn't even explore this issue, it scares me," Hunt said.

The "more than fifty organisations" which signed the petition clearly believe they are doing the right thing; they clearly believe that they are acting on behalf of all of the wonderful life-forms that have arisen on the Earth. How wrong they are! How deluded! How 'behind the curve'! In the future they will see the error of their ways. I am fairly sure that their actions will not have tragic consequences for the future of life on Earth (their actions won't lead to the extinction of complex life on Earth, and ultimately the extinction of life on Earth). However, if they could come to embrace the *New View of Humans in the Cosmos* that I am outlining in this book, then we could move forwards more quickly, and speed is of the essence. The sooner that we can learn how to effectively fulfil our purpose, the better it will be for both life on Earth and for ourselves as a species. If we are held back because of the view expressed by the "more than fifty organisations" then this will lead to great harm and suffering which could have been prevented; harm and suffering through events such as rising sea levels, climate change and extreme weather events. As, Professor Hunt realises: these organisations, and their views, are "scary". Let us embrace a *New View of Humans in the Cosmos*.

Technology

The phenomenon of technology is obviously central to my philosophical worldview. The human species *is* that part of life on Earth which has become technological, and it is this fact that distinguishes the human species from the rest of life on Earth. In other words, it is this fact that elevates the human species to a position of 'superiority'.

In this book, and in previous books, I have outlined why it is obvious to me that technology is in the interests of life on Earth; that is to say, I have outlined why it is obvious to me that the bringing forth of technology is a positive event in the evolution of the planet. This view does not entail that *all* aspects of technology are positive/beneficial/in the interests of life on Earth. Indeed, such a view would be blatantly absurd. Everyone knows that technology has both positive and negative aspects. Yet, when it comes to the environment, it is surprisingly common for people to have a polarised view of technology. At one extreme, the neo-Luddites see technology as a wholly bad phenomenon which is the cause of the environmental crisis. At the other extreme is the view that whatever environmental problems arise, technology can provide the solution; technology is conceptualised as a 'silver bullet', a panacea.

Sometimes people mistakenly believe that the 'silver bullet' view of technology is part of my philosophical worldview. The reason for this mistaken belief seems to be the conflation of two other beliefs: 1) The belief that the evolution of technology is a positive event for life on Earth; 2) The belief that technology can provide a solution for every environmental problem. These are two very different beliefs; I have the first belief and reject the second.

In *Chapter Eight* I outlined the ways in which technology is obviously in the interests of life on Earth. I also outlined, in *Chapter Seven,* the two different categories of environmental problems that exist. The second category is defined by the possibility of having either a technological or a non-technological solution. A serious view of the place of technology on an evolving planet needs to do this, it needs to embrace both the positive and the negative aspects of technology, and it needs to take into account both the immense power and the limitations of technology. Technology simultaneously eliminates risks which threaten the survival of life on Earth, and presents a danger to parts of life on Earth. A serious approach to the environmental crisis needs to outline which areas require technological solutions and which areas might be best dealt with through non-technological solutions. The extremes of the neo-Luddite view and the silver bullet view are equally useless.

The suffering caused by technology is crystal clear for all to see. When one hears on the news, as one regularly does, that a bus full of children has crashed and that a great number of them have died, what is one to think? One is likely to think that if it wasn't for the technology of cars/buses/etc. that this death and suffering wouldn't have occurred. One wonders: Why on earth do we whizz around at 80 miles per hour in bits of metal in confined spaces? Of course, the death and suffering which results from such whizzing about is a wholly negative effect of technology, but it is a price to be paid for the benefits of technology for life on Earth. Non-human animals also suffer immensely because of technology. Hit by cars, ripped to shreds by airplane propellers, shot by guns, harpooned in the ocean, killed by wind turbines. The suffering is there for all to see, a price to be paid for the overall benefits of technology for life on Earth.

There are good grounds for believing that the suffering caused by technology will reduce in the future, whilst the benefits of technology will become exceedingly clear. When this happens a widespread reconceptualisation will be possible – the human species, as that part of life on Earth which has brought forth technology, will be seen to be the saviour of life on Earth. The initial epoch of technological development – the bringing forth and proliferation of technology on a planet – is the most dangerous stage. As time progresses, technology can be more effectively controlled and suffering can be greatly reduced. For example, in the future the amount of road deaths looks set to be slashed as technological advancements enable cars to be self-driven. *The Metro* newspaper reported on 30 May 2012 that (*'Convoy of driverless cars completes 200km test run'*):

> a convoy of self-driven cars took to a public road for the first time. The convoy of Volvos kept a gap of 6m (19ft) as they travelled at speeds up to 52mph for 200km (124 miles) on a road outside Barcelona. The test cars were fitted with cameras, radars and laser sensors that allowed them to maintain a gap as they copied a lorry, which was controlled by a professional driver.

Clearly, the suffering resulting from the bringing forth of mechanised vehicles and roads initially increases as they spread over the Earth. However, there becomes a tipping point, and after this point technological advances enable the suffering to dramatically fall. In this realm, and in many others, technological solutions ultimately reduce the suffering which was previously caused by technology.

Whilst there are grounds for optimism, it needs to be remembered that technology is a complex phenomenon. In some areas technological development provides large benefits for life and no risk or suffering whatsoever (when the realm is considered in isolation); for example, the technology to protect life from an asteroid strike. In other areas, such as geoengineering the temperature of the atmosphere, there is an absolute need for the technology, but there is also risk and a potential for suffering. In this realm, one naturally hopes that the technology will work perfectly when it is first deployed. But, if not, then as with the evolution of technology in the realm of road safety, there is every reason to expect speedy progression. In other words, there might well be lots of small-scale geoengineering experiments which don't go exactly as planned, but if this happens then these experiments will lay the foundations for a successful technological solution which provides vast benefits and has negligible deleterious impacts.

Human Population & the Environmental Crisis

I have had several encounters with people who think that the human population should be restricted or reduced 'for the sake of the planet'. Anyone who has taken a course in Environmental Studies will be aware of the basic Environmental Impact formula:

$$EI = P \times R \times T$$

In other words, the total environmental impact of humans on the planet (EI) arises from a combination of three factors – the number of humans on the planet (P), the average per capita level of resource use (R) and the technological efficiency of producing these resources (T).

My encounters got me thinking about the nature of this formula. It is a very simple formula, if P increases then EI increases, if R increases then EI increases, and if T decreases then EI increases. However, behind the simplicity lurk a number of assumptions and complications. The primary assumption is that any increase in EI is bad. The primary complication concerns the nature of EI. What exactly is EI?

Let us first consider the primary assumption that any increase in EI is bad. This means that any increase in P or R, or any decrease in T, is bad. Now, one might believe that in the distant past, when P was very low, that an increase in P was not bad. If one believes this, then one is surely correct. However, it is important to recognise that in the distant past the EI formula did not exist. The very fact that the EI formula was devised indicates that humans see increases in P and R as a problem, as a bad/dangerous thing. So, I am suggesting that the assumption lying behind this formula has always been that any increase in EI (any increase in P or R, or any decrease in T) is bad; this seems to be intrinsic to the formula. Have you ever heard anyone say any of the following?

> For the sake of the planet I think that the human population size should be increased.
>
> For the sake of the planet I think that all humans should consume more resources.
>
> For the sake of the planet I think that we should use resources less efficiently.

I doubt it! The assumption is always that 'for the sake of the planet' equates to reductions in EI, not increases in EI. Do most humans really know what is in the interests of the planet? Or, do they just assume that they know? When everyone (or almost everyone) unquestioningly assumes something to be true, then there is a reasonable chance that they are wrong! Indeed, from the perspective of my philosophy, the almost continual increase in EI from the bringing forth of the human species to the current day, can be seen as a good thing. This continually increasing EI is effectively a direct measure of the growing strength of *the force to environmental destruction*. And, as we have seen, the growing strength of this force is a good thing, a sign that life on Earth is thriving and heading towards a successful technological birth.

Let us now consider the nature of EI. It doesn't seem to be a very useful thing to imagine. EI is an attempt to envision the total environmental impact of all humans that live on the Earth; it is a formula, an image constructed in the human mind. 'Out there', on the planet itself, there are simply a diverse range of individual environmental impacts. Believing that any reduction in EI (this creation of the human mind) is a good thing has some seemingly unsavoury implications. For example:

1. If 50 humans die in a motorway pileup this is good, as reductions in P reduce EI.

2. If unemployed people are put to work using simple technology, in the process replacing technologically more efficient machinery which requires no human labour, then this is bad, as reductions in T increase EI.

3. If 100 people move out of extreme poverty/near starvation and start eating more food this is bad, as increases in R increase EI.

4. If 100 people start eating less food and move into a state of near starvation this is good, as decreases in R decrease EI.

I am not convinced that we should see any of these things (these changes in EI) as being either good or bad for the environment. I am not convinced that these supposedly 'good' eventualities (the reductions in EI) would be 'for the sake of the planet'/beneficial outcomes for life on Earth. I think we would be much better off simply looking at individual situations/problems and giving up on the EI formula. When we do this then we can give up the simple idea that increases in P, or increases in R, are automatically bad.

For example, one major environmental problem is biodiversity loss. If P or R increase in a particular location this might lead to a serious loss of biodiversity. However, if the same increase in P or R occurred in another location there might actually be a resulting increase in biodiversity. The realities of the state of the planet cannot be adequately captured in a simple formula.

Another major environmental problem is climate change. I am sure you can easily appreciate that P and R can increase without having any impact on the climate; this means that EI would be increasing but that this wouldn't affect this environmental problem (although there could be a tipping point past which an increase in EI would have an affect).

Then, of course, there is human-induced global warming. Let us look at this environmental problem from the perspective of the EI formula. The assumption is that an increase in P (or an increase in R; or a decrease in T) is bad because it contributes to global warming. So, 'for the sake of the planet' P should be reduced, or maintained, or restricted in its growth. You will be aware that I believe that a proper assessment of the situation that we are in reveals that this environmental problem needs a particular solution; this solution is the active technological regulation of the GMST. So, this means that maintaining P, or restricting the growth of P, or reducing P, is not a solution to the problem. Future changes in P are of minimal significance to either the problem or the solution to the problem. The EI formula effectively becomes redundant. Future changes in P, R and T will affect EI, but they will not affect this environmental problem, or its solution, in any meaningful way.

So, the EI formula doesn't seem to be that useful, and it also leads to simplistic broad-brush thinking. For example, 'for the sake of the planet we should reduce P'. You might well recall, following what was said in *Chapter Ten*, that an outcome of my philosophy is that the best interests of life on Earth/the Solar System/the Universe are actually served by an increase in P (all reductions in P are bad).

The Growing Realisation of the Need for Geoengineering the GMST

In the article *The Need for a New View of Humans in the Cosmos* I cited evidence of the growing realisation that we need to geoengineer the temperature of the atmosphere; merely reducing emissions not being an option that is capable of stabilising the temperature of the atmosphere. In my overall philosophy this growing realisation has a special place in the cultural trajectory of human civilization; it is the stimulus which causes the human species to fulfil its purpose. It is this realisation which causes the human species to start actively technologically regulating the temperature of the Earth's atmosphere (the GMST). As I outlined in this article, we are currently living at a time when the realisation of the need for such technological regulation is growing, but the outcome itself is seen as an undesirable 'last resort'. In the future, the outcome will be seen not as an undesirable 'last resort' but as an inevitable requirement for the continued flourishing and existence of life on Earth. In this article I would like to briefly present yet more evidence of the growing realisation of the need for the active human technological regulation of the GMST.

Klaus Lackner and his colleagues at the Lenfest Center (part of the Earth Institute) in a paper published in the journal *Proceedings of the National Academy of Sciences* claim that there is a vital need for pulling CO_2 out of the atmosphere in order to regulate the GMST. They claim that such carbon capture and storage is the only solution to the situation that we face. In their paper, *Urgency of development of CO_2 capture from ambient air* (27 July 2012), Lackner and colleagues claim:

> In a way, it's too late to argue that we shouldn't consider [such] solutions. The concern that this kind of technology would give us an excuse not to do anything [to reduce carbon emissions] is wrong, because we're too late for that... We have to push very hard right now, and we have to have every means at our disposal to solve this problem.

I obviously agree with this statement, particularly that "we're too late for that". The mechanisms underpinning the trajectory of cultural evolution ensure that by the time humans realise what they have done, the only option, the only solution, is technological regulation of the GMST. If these mechanisms were different, and technological regulation could be avoided, then life on Earth would be doomed. Life on Earth needs technological regulation of the GMST, and this is why the biological-cultural trajectory of life ensures that "we're too late for that". This is how life gets what it needs in order to continue surviving and thriving.

Humans and Other Animals

I would like to say a little more concerning the following:

    1 The way that humans see their place in the cosmos

    2 The place of humans in the cosmos

Let us start with the first of these. Most humans don't seem to spend too much of their lives pondering the question of whether the human species has a special place in the cosmos. However, the question of *how the human species relates to the non-human animals which inhabit the Earth* seems to be a question which is more widely pondered. And these two questions are, in effect, the same. So, if humans think of themselves as simply one species of animal among many, then they are, simultaneously, subscribing to the view that humans do not have a special place in the cosmos. Whereas, if humans think that the human species is 'special' in some way, distinguished from all of the other non-human animals which inhabit the Earth, then they are, in effect, subscribing to the view that humans have a special place in the cosmos.

Due to the spread of evolutionary thinking – the greater realisation of the fact that the human species evolved from non-human animals – there is an increasing tendency for humans to think of themselves as just 'one species of animal among many'. This way of thinking arises largely because the thinker is simply considering a limited and narrow range of facts: 1) humans evolved from non-human animals, leads to 2) humans are simply one species of animal among many. This entails: 3) humans do not have a special place in the cosmos.

When one broadens the range of facts that one considers then things get much more complicated. Evolutionary thinking is perfectly compatible with the view that the human species has a special place in the cosmos. When one thinks of the cosmos as an evolving whole, giving rise to solar systems, planets, and life; when one starts to understand how life-bearing planets age; when one starts to comprehend the forces underpinning the biological-cultural evolutionary trajectory; then, one can see why the human species has a special place in the cosmos.

The view that the human species does not have a special place in the cosmos is also bolstered through a simple comparison of the attributes of humans with the attributes of non-human animals. This is how the view goes: *All animals have different attributes, so why should the human species (and the particular attributes of humans) be 'special/'superior'?* However, this way of thinking really won't do. The question is misplaced. The issue of whether the human species has a special place in the cosmos cannot be answered by simply comparing the attributes of humans with the attributes of non-human animals. The issue demands the consideration of a very broad range of factors, such as those mentioned in the previous paragraph.

The way that humans see the relationship between the human species and other animals goes through fads/phases –'non-special', 'special', 'non-special', 'special'. In contrast, the fact of whether the human species has a special place in an evolving cosmos is obviously unchanging.

It is exceedingly obvious to me that the human species has a special place in the cosmos, that it is not just one species of animal among many. This is the unchanging fact which undergirds the fads/phases. In the face of environmental destruction, this unchanging fact can surely be a source of some comfort. For, those who aren't aware of the bigger picture, those who just focus on the environmental destruction, are typically led solely to despair and frustration; they miss out on the underlying joy.

Animals Think like Humans

In my first book – *Is the Human Species Special?: Why human-induced global warming could be in the interests of life* (2010) – I outline how humans came to see themselves as fundamentally different to all of the other species of animals that inhabit, and have inhabited, the Earth. I also explain why all of the attempts that have traditionally been made to create a rational pedestal which elevates the human species to a position of superiority fail. In other words, for a great swathe of human history humans have been grasping around for an attribute which 'elevates' them above the rest of the animals which inhabit the Earth – an attribute which engenders human uniqueness – but all of these traditional attempts have been misplaced.

You will be aware of many of the attributes that have traditionally been assumed to 'elevate' humans to a position of superiority – possession of a soul, tool use, language, emotions, rationality, morality, thought, consciousness, self-awareness, culture, and so on. We are now living through an epoch in which it is slowly being realised that none of these attributes 'elevate' humans to a position of superiority. These attributes are possessed by many other non-human animals which inhabit the Earth; they cannot, in themselves, elevate the human species to a position of superiority (in ITHSS I make the case that *if* humans have a soul, then some non-human planetary life-forms also have a soul; alternatively, humans might not have a soul; either way, the having of a soul does not elevate the human species to a position of superiority).

A sign of this growing realisation is *The Cambridge Declaration on Consciousness*. In this declaration leading cognitive neuroscientists state that it is unequivocal that non-human animals are conscious and possess the cognitive ability to assess situations based on prior experience, and then act accordingly. The declaration was publicly proclaimed on the 7 July 2012 by Philip Low, David Edelman and Christof Koch. The declaration states that:

> the weight of evidence indicates that humans are not unique in possessing the neurological substrates that generate consciousness. Non-human animals, including all mammals and birds, and many other creatures, including octopuses, also possess these neurological substrates.
>
> (http://fcmconference.org/img/CambridgeDeclarationOnConsciousness.pdf)

I am glad that we are now entering the stage of planetary evolution in which we cease to see ourselves as being surrounded by 'inferior' animals with lesser attributes, and start to fully accept that there are no traditional unique 'superior-making' attributes which individual humans possess and that all non-human Earthly life-forms lack. This stage of planetary evolution is one in which necessary exploitation slowly gets replaced with care and compassion. It is also an era in which humans start to realise exactly why the human species has a special place on the Earth. The human species will soon come to fully embrace its role as the saviour of the life that has arisen on the Earth.

Earth 'Four Years from Disaster'

I thought I would share with you an article from *The Metro* newspaper (1 October 2012). On page nineteen there is a very small article entitled: *Earth 'four years from disaster'*. Here is the article in its entirety:

> FAILURE to address the threat of climate change is 'reckless and short-sighted', campaigners claim. Global policies have 'taken us backwards', with the world four years away from dangerous surges in temperature, a collection of dozens of charities and individuals say. The EU agreed rises should not exceed 2C (3.6F) if the worst impacts were to be avoided. Supporters including designer Dame Vivienne Westwood urged politicians to invest in a huge energy efficiency drive.

To summarise the article:

1   We are 4 years away from dangerous surges in the temperature of the Earth's atmosphere.

2   Lets deal with this scenario by urging our politicians to engage in energy efficiency savings.

It is hard to take this even remotely seriously. The situation we are in is assumed to be so dire that we are only 48 months from immense danger, and the proposed solution is to campaign politicians to make efficiency savings! The nature of the twin time-lags – the time-lags of the biogeochemical cycles of the Earth, and the time-lags inherent in cultural/political change, combined with the reality of the societal situation we face (the push for growth in recession-hit 'developed' economies, the push for growth in 'developing' countries, individual human motivation/desire), mean that this is nonsense. If we are 48 months from immense danger, then the solution is not to urge politicians to make efficiency savings! If we are 48 months from immense danger then the immense forces which are propelling us towards disaster are not going to be stopped by such a futile campaign.

Hope is an admirable attribute. However, if as one gazes up at the night sky one puts a straw into ones mouth and asserts *I am hopeful that I can blow the moon out of its orbit*, then one should not be taken seriously. One can be admired for one's ambition, but when the realities of the situation are explained to one then one will come to understand why one's 'moon-blowing efforts' were greeted with laughter.

Clearly, if we are 48 months from dangerous surges in temperature, due to the immense forces that have been gradually unfolding on the planet for thousands of years (*the force to environmental destruction*), then our only hope is to put the maximum amount of resources possible into our attempts to technologically control the temperature of the atmosphere. The forces which are still building up, the forces which are lurching us towards dangerous future temperature surges, can only be offset by an opposing matching force. If we are able to master the mechanics of technologically regulating the temperature of the atmosphere then dangerous temperature surges can be averted.

I will assume that you want to help the human species and that you want to help the rest of the life-forms which we currently share the planet with. I will assume that you want to prevent dangerous surges in temperature which would threaten humans and our non-human companions. I will assume that you believe that we are 48 months from dangerous temperature surges in the planetary atmosphere. What should you do to help the situation?

Don't fool yourself into believing that your time would be well spent "urging politicians to make efficiency savings". This would be a complete waste of your time. Even if you were successful, and the politicians made some efficiency savings, these savings would be swamped by increasing global carbon dioxide emissions over the next 48 months. Besides, the time-lags in the biogeochemical cycles mean that the atmospheric temperature over the next 48 months is largely determined by events that have already happened (in the absence of future technological regulation).

If you want to do something of real value, rather than waste your time, then spend every waking moment urging politicians, environmentalists, academics, the media, charities, and everyone you meet, that the need for full-scale geoengineering of the temperature of the Earth's atmosphere is real and imminent. You might initially be met with resistance, even derision, but in the future you will be looked back on as a hero.

The Futility of Emissions Cuts

In the last article, in response to an article in *The Metro*, I claimed that if global warming is an imminent threat that attempting to deal with it by making efficiency/resource savings is a hopeless strategy. After I wrote this article a paper was published by scientists Dr Jasper Knight and Dr Stephan Harrison in which they claim that it is too late to deal with global warming through emissions cuts. This entails that attempting to deal with global warming by making efficiency/resource savings is a hopeless strategy. Knight and Harrison state that:

> At present, governments' attempts to limit greenhouse-gas emissions through carbon cap-and-trade schemes and to promote renewable and sustainable energy sources are probably too late to arrest the inevitable trend of global warming
>
> (Nature Climate Change, *'The impacts of climate change on terrestrial Earth surface systems'*, Dr Jasper Knight and Dr Stephan Harrison, 14 October 2012)

I am glad that Knight and Harrison can see the futility of emissions cuts. Given this futility, what do they suggest that we should do? They argue for a switch in concentration from mitigation policies to adaptation policies. They believe that this switch will enable, in the face of global warming, optimal outcomes to be attained in terms of sustainability, food security and biodiversity. They claim that much more research is needed into how global warming will impact on Earth surface systems so that these optimal outcomes can be attained:

> Earth surface systems provide water and soil resources, sustain ecosystem services and strongly influence biogeochemical climate feedbacks in ways that are as yet uncertain.

When one appreciates that there are forces which are hurtling the planet towards future global warming, forces which themselves cannot be stopped, then it is perhaps understandable to conclude, as Knight and Harrison do, that *we cannot stop these forces/global warming, so let's adapt to global warming.*

However, I find this conclusion a deeply troubling one; troubling because it is clear to me that we are not talking about just a small degree of global warming. If in the future there was only a small amount of global warming, then Knight and Harrison might be spot on in their urgings for adaptation policies to attain optimal outcomes for food security, sustainability and biodiversity.

There is obviously a wide range of views, and a great degree of uncertainty, concerning exactly how much global warming can be expected in the near future. My interpretation of the way that the biogeochemical cycles of the Earth have been 'temporarily temporally perturbed' (chiefly though the temporary storage of carbon in the thermohaline), viewed against the backdrop of the weakening planetary homeostatic regulatory capacity, leads me to conclude that there are time-lag forces which are set to unleash a massive amount of global warming in the near future. This amount is of such a magnitude that it could wreak devastation on the human species and most of our fellow planetary companions.

If this is right, then whilst Knight and Harrison are sensible to call for a greater emphasis on adaptation policy research, we should fully embrace the reality that adaptation policies themselves cannot adequately deal with anything other than a trifling amount of global warming. The policy we need is not adaptation, it is technological regulation. We need to offset the forces which are set to cause global warming, not adapt to global warming. In other words, we need to use technology to actively regulate the atmospheric temperature in order to stop global warming from occurring. Such technological regulation needs to be our main response to global warming. There is a secondary need, a much lesser but still important need, for the adaptation policies which are championed by Knight and Harrison.

Prepare for Extreme Global Warming

If you have read the book up to this point then you will be aware that there are three key aspects to my view of the place of humans in the cosmos:

1) The human species will (well, it already has) 'inevitably' set in motion the forces which will potentially cause very severe global warming. This was 'inevitable' from the moment that life first arose on the Earth.

2) This is not a bad thing. This is because it is a side-effect of the planet giving birth to the technological protection which benefits life on Earth. Technological protection takes many forms, but the most important one is the ability of life to technologically regulate the temperature of the atmosphere. Without this ability the planet will inevitably fall back to a state of lifelessness. So, 1) above, and in particular, concerns about 1), will lead to the human species technologically regulating the temperature of the Earth's atmosphere; this regulation will significantly bolster the probability that life will continue to exist in this region of the cosmos in the future.

3) The popular view that *life on the planet has existed for millions of years without humans, that humans are the destroyers of life, and that life would survive and thrive on the planet well into the distant future if humans were to go extinct* is wrong. This view is simplistic and plain wrong. The human species is the saviour of life on Earth.

Of course, there are those who disagree with everything that I have just said. In several of the preceding articles my focus has been on convincing you of the truth of 1). In the rest of this article I will provide a little more support for the truth of 1). A recent study has concluded that most predictions concerning the amount of global warming that we can expect on the planet by the end of the century are way too low. The authors claim that we should expect what are currently "the most extreme predictions" (the most massive amount of warming) to prevail. This implies that new predictions are needed; predictions which are much more extreme than the present most extreme ones!

John Fasullo, of the National Center for Atmospheric Research in Boulder, summarises the report as follows:

> Warming is likely to be on the high side of the projections
>
> (The Washington Post, 'Warmer Still: Extreme climate predictions appear most accurate, report says', Brian Vastag, 8 November 2012)

The current extreme projections are for a "devastating" increase of eight degrees Fahrenheit by 2100. According to *The Washington Post* article:

Such an increase would substantially overshoot what the world's leaders have identified as the threshold for triggering catastrophic consequences. In 2009, heads of state agreed to try to limit warming to 3.6 degrees, and many countries want a tighter limit.

It is surely only a matter of time before everyone – environmentalists, climate scientists, politicians, everyone on the planet – comes to realise the truth of 1), and also comes to realise that the active technological control of the temperature of the Earth's atmosphere is the only feasible solution to the situation that we face. The sooner that this day of realisation arrives the better it will be for all of the life-forms that currently inhabit the Earth.

Emissions Cuts: The Gap between Ambition & Reality

In the previous few articles I have been considering greenhouse gas emissions. I have cited journal and newspaper articles which reveal a growing realisation that greenhouse gas emissions are at a level which makes it very likely that (in the absence of geoengineering) there is going to be an extremely dangerous increase in the temperature of the planetary atmosphere later this century.

Despite the increasingly widespread realisation that this is so, there is still optimism that this extremely dangerous scenario can be averted if governments get together and agree to significantly reduce their emissions. One is tempted to believe that many people are blinkered into thinking that the only real or sensible solution to the global warming problem is to significantly reduce emissions. So, whilst the reality is that this isn't going to happen in the required timescale, people still want to believe that it can happen (and, of course, given that the damage has already been done, this entire endeavour is a waste of time). People are, in the main, optimistic; so, if there is thought to be only one solution to a problem, then it is natural to keep on hoping that this solution can be attained, whatever the reality of the situation. There is clearly a gap between ambition and reality. Indeed, in the recently published *Emissions Gap Report 2012*, Achm Steiner, the executive director of the United Nations Environment Programme, states that the report:

> provides a sobering assessment of the gulf between ambition and reality

In the report, the reality of emissions being way too high to prevent dangerous warming is accepted; this *reality* is completely out of kilter with the *ambition* to have massively lower emissions. Yet, the report still tries to be optimistic in its conclusions, stating that it is "technically possible" that emissions can be slashed and that dangerous warming can be averted.

What does it mean for something to be "technically possible"? This way of looking at the situation seems to simply be a case of misplaced optimism and it could be extremely dangerous; it just seems to mean not theoretically impossible. It is also "technically possible" that I could win the National Lottery Jackpot every week for a year. However, when one gets serious, one needs to leave these theoretical "technical possibilities" aside, and return to reality! The reality is that *the force to environmental destruction* will continue to dominate; greenhouse gas emissions will not be slashed in the timeframe that is believed to be required (it is actually too late anyway, because as we saw in *Chapter Nine* the damage has already been done; so, the immediate global slashing of emissions actually wouldn't change the situation we face).

Another report has recently been released by The World Bank entitled *'Turn Down the Heat: Why a 4 Degree Warmer World Must Be Avoided'* (2012):

http://climatechange.worldbank.org/content/climate-change-report-warns-dramatically-warmer-world-century

According to this report:

> the world is on track to a "4°C world" marked by extreme heat-waves and life-threatening sea level rise.
>
> As global warming approaches and exceeds 2°C, there is a risk of triggering nonlinear tipping elements. Examples include the disintegration of the West Antarctic ice sheet leading to more rapid sea-level rise, or large-scale Amazon dieback drastically affecting ecosystems, rivers, agriculture, energy production, and livelihoods. This would further add to 21st-century global warming and impact entire continents.
>
> The projected 4°C warming simply must not be allowed to occur—the heat must be turned down. Only early, cooperative, international actions can make that happen.

In this report there is clearly an admirable ambition to avoid the extreme danger that we face from a massive increase in temperatures triggered by an above 2°C rise (due to the associated nonlinear tipping points which are likely to result in a runaway warming effect). Yet this ambition is again out of kilter with reality. This is because there is a misplaced belief that the desired outcome can be achieved through emissions cuts.

The sooner that reality is fully accepted the better. When one realises that one's proposed solution to a problem is completely out of kilter with reality, then it is usually best to seek another solution. This isn't a bad thing, a sign of failure. In this case, the realisation will simply cause one to shift one's energy and focus to a real solution to the problem; a solution which is in accordance with reality. Of course, you know what this solution is: the geoengineering of the temperature of the atmosphere.

As a final note, I should reemphasise that many people still seem to believe that if emissions were 'magically' slashed from tomorrow, that everything would be fine. However, this seems to be another case of the widespread human need to be optimistic; there are good reasons to believe that in reality this would not make any difference; the damage has already been done due to past events whose time-lag biogeochemical perturbation effects have yet to be manifested in increasing atmospheric temperatures.

The World Bank report is right that "only early, cooperative, international actions" can avoid a 4°C rise. The real question which needs to be addressed is what these actions actually are. When reality is accepted, then the ambition to avoid such a rise can be met with a solution that can work (technological regulation of the atmospheric temperature), rather than with a completely ineffectual alternative (attempting to slash emissions). So, if reality can be widely accepted then there is still reason to be optimistic; one just needs to place one's optimism in the right solution.

Accelerating Polar Ice Melting & Geoengineering

I came across an interesting article in the *i* newspaper which highlights the accelerating rate of polar ice melting (*'Polar ice melting three times as fast as 20 years ago'*, Lewis Smith, 30 November 2012, p. 25). The article states that:

> More than 4,200 gigatones were lost from the polar ice sheets from 1992 to 2011, an average of 223Gt a year and rising. Researchers described the rate of losses as being at "the very upper end" of forecasts published by the Intergovernmental Panel on Climate Change (IPCC) in 2007.
>
> During the 1990s ice sheet loss accounted for 10 per cent of sea level rises, but in the last five years it has risen to more than 30 per cent, the researchers said.
>
> The Antarctic ice sheet contains 30 million cubic kilometres of ice and holds around 90 per cent of all the fresh water on the surface of the Earth. If the whole Antarctic ice sheet melted, sea levels would rise by more than 60 metres.

What are we to make of this? I think we need a three-pronged approach:

1. Accept our limitations

2. Utilise our strengths

3. Properly acknowledge the level of uncertainty about what might happen in the future

Let us consider each of these in turn. Accepting our limitations is perhaps the most important of these three factors. An exploration of our limitations requires a consideration of two different types of forces which are at work on the Earth (these two forces are different aspects of *the singular force* which propels the evolution of the Earth/Solar System; in previous parts of this book I have conceptually divided this singular force into *the force to environmental destruction* and *the force to environmental sustainability.* The two types of forces we are about to consider below are both part of both *the force to environmental destruction* and *the force to environmental sustainability*):

> A) Biogeochemical Forces – The biogeochemical forces are the trajectories inherent in the biogeochemical cycles of the Earth. These forces operate over a variety of timescales, but many of the important cycles operate over relatively long timescales; these cycles are characterised by inertia and by lengthy time-lags between cause and effect. These cycles have been slowly building in momentum along a particular trajectory over hundreds and thousands of years. The result of this trajectory is the current changes we are seeing on the planet, changes such as a worrying acceleration in polar ice melting.

B) Socio/Economic/Political/Cultural/Individual Forces – These are the forces which exist in individual humans and in the frameworks and structures which humans have created. The forces in this domain have also been evolving, feeding off each other, and slowly building up for thousands of years. These forces have propelled cultural evolution towards globalisation and the human technological modification of large parts of the planet. In resulting in this modification, these forces have obviously affected the biogeochemical cycles of the Earth, they have served as an 'input' which has exacerbated, altered, or initiated, biogeochemical trajectories; in this way these forces have contributed to outcomes such as the acceleration in polar ice melting.

These two types of forces have trajectories which are not about to radically change; they are characterised by inertia, by strength, by time-lag effects, and by a gradual rate of change. To explore the nature of these two forces let us engage in a thought experiment. Let us imagine that the Socio/Economic/Political/Cultural/Individual Forces immediately ceased; this entails that there would be no more human 'interference' with the planet (no construction, no greenhouse gas emissions, no car or airplane travel, no deforestation, etc.). In this scenario what would happen to the Biogeochemical Forces, particularly in the realm of polar ice? The polar ice would still continue to melt at an accelerating rate for the foreseeable future, and atmospheric greenhouse gas concentrations would also continue to increase for the foreseeable future. What this means is that whilst the two types of forces are intricately interconnected, they are also largely independent when they pick up their own momentum.

Imagine a large boulder on top of a hill; it is just sitting there until a human pushes it, after which it starts rolling. The human obviously causes it to start rolling, but once rolling the human cannot stop it, it 'inevitably' keeps on rolling until it gets to the bottom of the hill! Once it has started to roll, the movement of the boulder to the bottom of the hill can be thought of as a self-propelling force. These self-propelling forces exist in both A) and B) and they can be catalysed by A) forces/events, B) forces/events and other forces/events. The movements of these self-propelling forces can also lead to the passing of a biogeochemical tipping point which catalyses a chain reaction of future consequences; for example, the melting of the whole Antarctic ice sheet, or runaway global warming.

We need to accept the reality of the situation. We need to accept the existence, the strength, the inertia, of these two forces. We need to accept our inability to stop these forces. For the foreseeable future humans are going to keep on consuming, modifying and altering. For the foreseeable future the Antarctic ice is going to continue to melt, and greenhouse gas emissions are going to continue increasing. We are in peril, and life on Earth is in peril. We really need to utilise our strengths.

What are our strengths? This is obvious. We have walked on the moon, sent probes to Mars, and created a feast of engineering delights across the planet which would have utterly bewildered our ancestors. Surely, with all of that engineering expertise, pulling some carbon back out of the atmosphere should be a very simple affair. Releasing carbon from its underground storage areas and moving it into the atmosphere was a fairly simple human activity which we have already achieved; the reverse movement is surely not beyond us. This reverse movement is all that is required in the immediate future. We just need to realise the urgency of the need. Unfortunately, there are many powerful voices which are 'emission-reduction obsessed'; they do not even see the need, so they definitely do not see the urgency of the need! These voices are unwittingly trying to pilot us towards a very dark future.

Finally, let us consider the existence of uncertainty. There is obviously a very high level of uncertainty concerning the future. Forecasts concerning the future states of the biogeochemical cycles have been made in the past, and as time has passed the outcomes have often been at the very worrying end of forecasts (the most extreme changes, as with polar ice melting). As the two types of forces that we have considered continue on their trajectory there is a good chance that the changes which occur to the states of the biogeochemical cycles will far exceed the most extreme forecasts that are currently being made. Whilst there is uncertainty, all the danger is on the upside. In other words, the most extreme forecasts could be overshot by a small amount, a large amount, or an exceptionally massive amount! In the face of all this uncertainty we most definitely need our most precious gift at our disposal, our exceptional ability to modify, to geoengineer.

So, we need to accept our limitations, utilise our strengths, and properly acknowledge the level of uncertainty that we face. This seems like a sensible strategy for the future. All of these factors lead to the conclusion that we need to overcome the current 'emissions-reduction fixation' and push ahead with geoengineering efforts in the immediate future. Our future, and the future of life on Earth, depends on it.

Evolution versus Creationism

In the last few articles I have concentrated on geoengineering. I have considered both the need for geoengineering and why such an outcome would be a positive event for life on Earth. This need and outcome are firmly embedded within a view of the universe, the Solar System, and life on Earth, as evolving entities. Given this, I was slightly disturbed by some of the anti-evolution views which were expressed in *The Metro* newspaper and I wanted to comment on these views here. The following views appear under the heading: *'Gaps in our fossil records only open door to attacks on Darwin's Theory'* (12 December 2012, p. 14):

> Creationists accept variation within a kind (i.e. within the dog 'family', cat 'family', bovine 'family', equine 'family', etc.) but reject the notion of one type of creature, such as a wolf, turning into a completely different creature, such as a whale.

> the fossil record contains exactly what you would expect to find if the biblical account of creation were true.

> As a creationist, I agree there is such a thing as micro-evolution (changes within species). However, the fossil record does not support macro-evolution, which claims all species are related to each other and, for example, that we

are related to apes and descended from fish. Tens of millions of fossils have been dug up and still there is not a single clear, undisputed case of a 'missing link' between species. The fossil record consistently supports creation of separate species, not gradual evolution from micro-organisms to humans.

The people who made these comments seem to believe that if Charles Darwin's proposed mechanism of evolution – natural selection – is false, then this is a good reason to reject evolution and embrace creationism. In other words, these people take the gaps in the fossil record to be 'evidence' that natural selection is false and therefore conclude that creationism is true. This misunderstands Darwin's legacy. There are three meanings of evolution:

1. Evolution as Fact – species are not fixed but arise out of and develop into other species.
2. Evolution as Path – the actual routes that evolution has taken.
3. Evolution as Mechanism – the power that lies behind evolutionary change.

Darwin's achievement was to establish beyond reasonable doubt the truth of 1). He had very little to say about 2), but he did postulate a possible mechanism for 3) – natural selection. One should keep in mind that Darwin's legacy chiefly concerns 1). If his postulated evolutionary mechanism is partly, or wholly wrong, then this doesn't undermine 1). If one rejects Darwin's proposed evolutionary mechanism, as many evolutionists do, then one can propose a more plausible evolutionary mechanism that is in accordance with 1). One cannot reasonably reject 1) on the basis of rejecting a single possible mechanism.

The Metro quote above states that:

> The fossil record consistently supports creation of separate species, not gradual evolution from micro-organisms to humans.

The important word here is "not". In other words, the belief is being expressed that IF "separate species" came into existence THEN this entails that there was no "gradual evolution from micro-organisms to humans" (species were 'created' rather than 'evolved from other species'). Of course, there is actually no such entailment. If one rejects "gradual evolution from micro-organisms to humans" then one is rejecting 1) above. However, such a rejection does not follow from an acceptance that "separate species" came into existence. This is because there are very plausible evolutionary mechanisms which entail that "separate species" came into existence. In other words, one can believe, like I do, that there is "gradual evolution from micro-organisms to humans" AND that "separate species" come into existence throughout the evolutionary process.

In their book *Acquiring Genomes* (2002) Lynn Margulis and Dorion Sagan propose one such evolutionary mechanism with their symbiogenetic theory of animal speciation, as we saw in *Chapter Ten*. In my book *An Evolutionary Perspective on the Relationship Between Humans and their Surroundings: Geoengineering, the purpose of life & the nature of the universe* (2012) I provide a detailed exploration and defence of the symbiogenetic theory of animal speciation. In this book I also explore various theories of biological evolution, the relationship between biological evolution and non-biological evolution, and the extent of our possible knowledge concerning the mechanisms of evolution.

Those who jump straight from a belief in "separate species" coming into existence to *creationism* don't seem to understand that there are evolutionary mechanisms which entail "separate species" coming into existence. It is really useful to keep in mind the three meanings of evolution outlined above. If one does this then one can see that doubts about 2) – the fossil record and evolution as path – do not undermine 1). Furthermore, doubts about proposals for 3), such as Darwin's idea of natural selection, also do not undermine 1).

In short, the way that creationists resort to the fossil record to defend their view is flawed. Their critique of the data concerning *evolution as path* is built upon on an assumption concerning the importance of a particular evolutionary mechanism – natural selection. There is a mass of evidence for the reality of *evolution as fact*. There is much uncertainty concerning *evolution as path* and *evolution as mechanism*, but one cannot reasonably use this uncertainty to reject *evolution as fact* and embrace creationism.

The Calm before the Carbon Storm

There are three distinct positions when it comes to the human-induced global warming debate:

1. Human-induced global warming is happening but it is not a cause of immediate grave concern

2. Human-induced global warming is happening and it is a cause of immediate grave concern

3. Human-induced global warming is not happening

I find it worrying that the various advocates of these three positions tend to focus their debate on very short-term factors. The debate typically focuses on considerations such as:

A. Has there been an increase in extreme weather events due to human-induced global warming?

B. Was the massive hurricane last month due to human-induced global warming?

C   How much of recent temperature changes are due to 'natural variation' and how much are due to 'human-induced' effects?

D   Is the atmospheric temperature likely to rise over the next 5/10/20 years due to human-induced global warming?

These questions have one thing in common. They focus on the present, the immediate past, or the very near future. The problem is that this approach is quite inappropriate when it comes to the phenomenon of human-induced global warming. Why is this? I will try and briefly explain.

The brief answer is: time-lags and timescales.

A slightly elucidated answer:

The biogeochemical cycles of the Earth have been 'carbon perturbed' to a massive degree but the effects of this perturbation will not start to become manifest until the second half of this century.

In other words, all of the above questions (A-D) at best completely miss the point, and at worst are of trifling insignificance and are dangerous because they distract attention from the real global warming problem that we face.

A longer answer:

i) The atmospheric temperature of the Earth has been favourable for life for so much of the Earth's history because the increasing amount of solar radiation reaching the Earth from the Sun (non-human-induced global warming) has, until very recently, been offset by an increasing amount of carbon being stored under the surface of the Earth (as 'fossil fuels'). This storage is a key component of the Earth's homeostatic regulatory capacity.

ii) Very recently humans have upset this long-standing atmospheric regulatory process. The temperature of the Earth has only been favourable for life in the recent history of the planet because so much carbon had been stashed away under the surface of the Earth. Humans have released an enormous amount of this stored carbon and released it back into the biogeochemical cycles at the surface of the Earth. This means that all of this carbon can now potentially exert a very large warming effect on the atmosphere.

iii) The effects of this enormous release on the atmosphere are not immediate. A large part of the released carbon first enters the ocean and most of this finds its way into the deep ocean thermohaline circulation. Since the start of the Industrial Revolution most of the released carbon has simply entered this deep ocean circulation and barely any of it has re-emerged yet. The carbon is simply slowing moving through the deep ocean. A slowly ticking atmospheric time-bomb! It won't be until the second half of this century that this massive amount of carbon that humans have extracted from below the surface of the Earth will start emerging from these deep ocean currents. When this carbon starts to 'pour' out of the ocean a very significant increase in atmospheric temperature can be expected (if we are not technologically prepared).

iv) The question of whether *current/past/immediate future* climate events and temperature variations are 'natural' or 'human-induced' completely misses this bigger picture. The bigger picture is all about time-lags and longer timescales.

v) The climatic regulatory system [i) above] has been massively perturbed and the effects of this have not yet become manifest, and will not become manifest until the second half of this century (due to the very slow-moving pace of the thermohaline circulation).

So, when we realise all of this the question becomes:

What do we do?

Well, clearly, reducing carbon emissions is not going to solve the problem. Reducing current and future emissions, in itself, would be a good thing, but it is not going to solve the problem that we face. The massive perturbation will not be offset by slightly reducing the amount by which we continue to perturb the system! The only thing that we can do is to be as technologically prepared as possible. By the second half of this century we can be in a position which enables us to be technologically proficient at pulling carbon out of the atmosphere. This is the only way that we can restore the long-term atmospheric regulatory system [i) above] to a state of short-term balance. When the carbon starts gushing out of the ocean we can simply start pulling carbon out of the atmosphere.

In fact, the sensible option would obviously be to start pulling lots of carbon out of the atmosphere before the massive store of carbon in the thermohaline starts to gush into the atmosphere. This will enable us to perfect the technique, whilst also preventing other minor negative effects (such as extreme weather events) which might occur in the meantime, effects resulting from atmospheric greenhouse gas levels being higher than optimal.

The next question becomes:

Is all of this a 'good' or a 'bad' thing?

My reason for posing this question is to get you thinking about the even bigger picture. This technological solution will only put the long-term atmospheric regulatory system [i) above] back to a state of short-term balance. The problem is that this regulatory system cannot last forever because there comes a point when the carbon in the atmosphere effectively runs out! And, well before this point is reached, decreasing atmospheric greenhouse gas concentrations reach a level of 'lowness' that makes homeostatic regulation problematic; as we have explored in this book, this level of 'lowness' was reached on the Earth a long time ago. What this means is that the future existence of life on the planet requires a longer-term technological solution. The increasing solar output of the Sun needs to be directly intercepted before it reaches the Earth's atmosphere

(Path 3). Such an activity is required if humans and other 'complex' life-forms are to have a long-term future on the planet.

To summarise, given that the survival of life on the Earth is a good thing, this means that the development of the human capacity to technologically regulate the atmosphere in various ways (both pulling carbon out of the atmosphere and directly blocking incoming solar radiation) is a very good thing for life on Earth. The short-term human carbon perturbations are a side-effect of becoming technological and they are also the stimulus which catalyses the technological development which life on Earth needs in order to survive and thrive.

So, the answer to our question is:

All of this is an exceptionally good thing!

Perceptions of Global Warming

In the previous article I outlined my concern at the fact that much of the debate relating to human-induced global warming tends to focus on very short-term factors. This is a problem because the factors which are of importance are long-term rather than short-term.

In this article I will delve into this subject a little deeper. Firstly, I will mention a recent study carried out in the US relating to how attitudes to global warming are formed. Secondly, I will put this into a broader theoretical context by proposing that the way that the human perceptual processes work almost makes it inevitable that, whatever the subject, there will be a tendency for humans to draw conclusions on the basis of the short-term at the expense of the long-term.

So, to the US study. A study carried out at the University of British Columbia concluded that the local weather (particularly temperature) plays a major role in influencing public and media opinions on the reality of global warming. If there is a period of cold weather then there is a large increase in public and media scepticism towards global warming. Whereas, during a short-term hot spell there is much greater public and media concern about global warming. The findings of the study were published in 2013 in the journal Climatic Change ('*The influence of national temperature fluctuations on opinions about climate change in the US since 1990*', Donner S.D. and McDaniels J.). In the article Professor Simon Donner states that:

> Our study demonstrates just how much local weather can influence people's opinions on global warming... We find that, unfortunately, a cold winter is enough to make some people, including many newspaper editors and opinion leaders, doubt the overwhelming scientific consensus on the issue.

People are clearly heavily influenced by the present when they form their opinions about the future. One can too easily think: *If it is very cold now, how can global warming be a problem in the foreseeable future?* This way of forming opinions about issues which are long-term in nature, multifaceted, and time-lag dependent, is clearly hopeless. Long-term issues require a consideration of long-term factors and processes, not of transitory short-term fluctuations which exist in the present!

Why does this tendency for conclusions to be so heavily influenced by short-term factors exist? There are, I believe, good reasons for the seeming inability of the majority of humans to adequately account for the long-term in their thought processes. There are surely many factors which are of relevance here. However, I consider there to be two factors which are of greatest importance. Firstly, the human lifespan is itself only very short-term compared to long-term processes, such as the global warming processes, which humans think about. Secondly, the human perceptual apparatus operates in such a way as to reveal and highlight only very short-term movements (in contrast, the movements which are of importance in the global warming debate are very long-term movements). The result of this is that human thought processes tend to concentrate on these very same short-term movements (in other words, humans typically think about what they perceive). In the rest of this article I will give you a taste of these two factors. The following section is drawn from my book: *An Evolutionary Perspective on the Relationship Between Humans and their Surroundings: Geoengineering, the purpose of life & the nature of the universe* (2012).

The Inevitable Temporal Constraint:

The human perceptual apparatus is also inevitably constrained because it has in-built temporal constraints; it is only able to perceive movements from an exceptionally narrow temporal perspective. This is possibly a hard thing to envision; how is one to get a handle on this inevitable constraint? Let us start with the evolutionary perspective.

The universe has been evolving and moving in various ways for billions of years; in contrast, the average contemporary human will be lucky to reach the age of one hundred years. One can barely comprehend what it would be like to perceive a movement that spanned a thousand years or a million years – in contrast to the movements which our perceptual apparatus has evolved to perceive, such as the running of a wild animal towards one, which is a movement which lasts a matter of seconds – but such long-term movements clearly exist. The human perceptual apparatus has evolved to connect to short-term movements and is unable to connect to long-term movements.

How, exactly, does the exceptionally narrow temporal perspective from which humans are able to perceive movements in their surroundings inevitably constrain their perceptions? It is perhaps helpful to start by considering a relatively short-term movement which a human could, in principle, be able to perceive. It is possible that one could perceive the movement pattern that is the Earth taking 365 days to move around the Sun. If one was located in an appropriately positioned space station, and was able to continuously observe for 365 days, then one would be able to perceive this movement (of course, nearly all, if not all, humans alive at the moment would not be able to do this as they need to sleep roughly every 24 hours). When we start to consider slightly longer-term movement patterns, those that exceed the lifespan of a human, then it is obviously the case that it is impossible for a human to be able to observe these movement patterns (a human can only observe whilst they are alive!).

Why is this important? If a human only has perceptual access to a small temporal slice of a movement, then that human is not in a position to accurately judge the nature of that movement. The inability of humans to perceive long-term movements means that they are likely to conceive of much of their surroundings as mechanistic – this is because the small segments of movements that humans are able to access appear to them to be mechanistic. If humans had perceptual access over a longer temporal window then all of the movements which humans perceive in their surroundings might appear to them to be non-mechanistic. So, the inevitable temporal limits of the human perceptual apparatus can easily lead one to conceptualise the vast majority of one's surroundings as mechanistic – as very different from humans.

I will use an example to clarify this point. Let us consider a series of very short-term movements such as all the movements of the players on the pitch in a 90 minute football match. If you perceived this series of movements over a 90 minute period you would, no doubt, conclude that they were non-mechanistic. However, if you only had perceptual access to the first second of the match what would you conclude? The movements which you were able to perceive within this temporal window would not be of a sufficient duration for you to be able to conceive of them as non-mechanistic. You would, no doubt, conclude that the movements were mechanistic. It is only if you had a longer time slice of perceptual data that you would be able to conclude that the movement which you previously conceived to be mechanistic is actually part of a much longer duration movement pattern which you would now wish to assert is non-mechanistic.

I hope you can see this. The universe is the 90 minute football match. All of the perceptions that a human can have of the universe occupy the first second of the match. Humans form their conceptions of the universe based on this first second, but the universe isn't the first second, the universe is the whole 90 minute match! The human perceptual apparatus is clearly inevitably temporally constrained.

***

As I said, the above section is reproduced from another book in order to give you a taste of the inevitable temporal constraint inherent in the human perceptual apparatus. In this article we have seen that human thought processes tend to be dominated by what is happening in the present and by what humans are able to perceive. That which is long-term in nature is not apparent in the present and its nature can also be imperceptible to humans; this means that it typically gets ignored or glossed over. In this way, the factors of central importance in the global warming debate, which are long-term in nature, tend to get ignored or glossed over.

Global Warming: Perceptions, Responses & Energy Policy

An article that I came across in *The Sunday Telegraph* raises some questions which are related to the issues that we have just been considering. Here is the gist of the article:

> hidden in the small print of the Budget, were new figures for the fast-escalating tax the Government introduces next week on every ton of CO_2 emitted by fossil-fuel-powered stations, which will soon be adding billions of pounds to our electricity bills every year... the Coalition is... hell-bent on driving our... coal-and gas-fired plants out of business.
>
> So we are doomed to see Britain's lights go out, all because the feather-headed lunatics in charge of our energy policy still believe that they've got to do something to save the planet from that CO_2-induced global warming that this weekend has been covering much of the country up to a foot deep in snow. Meanwhile, the Indians are planning to build 455 new coal-fired power stations, which will add more CO_2 to the atmosphere of the planet every week than Britain emits in a year.
>
> (*'Chilly, isn't it? It's payback time for our energy policy'*, Christopher Booker, 24 March 2013, p. 34)

There are a couple of important points arising from this article:

1. The article provides further evidence of the unhelpful and misplaced media concentration on short-term factors (the weather today!) in the debate concerning global warming. Recall that in the last article I wrote that:

> A study carried out at the University of British Columbia concluded that the local weather (particularly temperature) plays a major role in influencing public and media opinions on the reality of global warming.

The Sunday Telegraph article is a very good example of this, attempting to discredit the reality of global warming because it is currently snowing in parts of England! One cannot take such a connection remotely seriously, but it is an example of the (problematic) way that the human mind works, overly concentrating on the very short-term, the immediately experienced reality.

2. The article highlights the continued strength of *the force to environmental destruction*. In the UK we might be facing future power cuts due to the attempt to reduce CO_2 emissions, but at the global level (and the broader UK level; e.g. the number of UK airplane flights) these efforts are dwarfed by the larger reality. As Booker says in the article:

the Indians are planning to build 455 new coal-fired power stations, which will add more CO_2 to the atmosphere of the planet every week than Britain emits in a year.

One might be troubled by this and one should be troubled by this. One might be environmentally aware and want to 'do the right thing' relating to the environment; one might try and live sustainably by minimising one's 'carbon footprint'; one might feel that one should make an effort. If so, one probably rationalises the situation as follows:

> *What I do might be miniscule in the bigger scheme of things (environmental changes at the planetary level) but if enough individuals act like me then it will make a significant difference.*

However, within one there is likely to be a nagging doubt, a sense of one's actions being ultimately futile, simply a gesture which makes one feel better about oneself *(I have purchased a reusable supermarket bag so I must be a good person!)*. At a deeper level the nagging doubt exists, the realisation that all that one has done is buy a shopping bag, whilst in India 455 coal-fired power stations are about to be built (just one example of the larger reality), and that one's lifestyle is still highly unsustainable (on the one hand a reusable shopping bag, on the other hand all of the supermarket packaging on the products that get put into the bag, the air miles of this food, the air miles of one's holidays, the car journey to the supermarket and to work, etc., etc.). One should realise that, in reality, one really is insignificant in the bigger scheme of things. One should realise that one is especially insignificant if one thinks that one's attempts to be environmentally-friendly make any difference whatsoever to global-warming at the planetary level.

There is a seemingly contrary view, a view which asserts that everything that everyone does is of significance. A good way to think of this view is through Chaos Theory and the 'butterfly effect'. According to this way of thinking, the universe is intimately interconnected to the extent that one tiny action, such as the flap of the wings of a solitary butterfly, can be the catalyst for a massive distantly-located change, such as the coming into existence of an enormous hurricane at the other side of the planet. In the realm of human action relating to the environment, it could be thought that the actions that a single human makes in reducing their carbon footprint could prevent a tipping point from being passed. The passing of the tipping point could, let us suppose, result in a runaway greenhouse effect; this means that every action that every human takes can certainly be thought of as potentially of great significance. And, in one sense, this is true. Individual humans acting in accordance with the inner drives and motivations that the universe has endowed them with, is the force which propels human civilization forwards. Every environmentally-friendly action taken by every individual, when aggregated, forms part of *the force to environmental sustainability*. This force is important, so the individual actions which constitute it are also important.

Can this seemingly contrary view be compatible with the view that one is insignificant in the bigger scheme of things? I think so. If an environmental tipping point is passed, this will certainly be due to a single action or event. However, when we consider the immensely powerful forces that are driving the evolution of the planet – biological evolution and cultural evolution – then we can see that this trajectory, which includes the passing of tipping point thresholds, will not be meaningfully altered by the actions of particular individuals. In other words, both *the force to environmental sustainability* and *the force to environmental destruction*, whilst being constituted by individual actions, are not in any meaningful way affected by individual actions. In other words, an individual lacks the ability to change the trajectory, or speed, with which these forces unfold and evolve.

A useful analogy might be to imagine that one is in the centre of a thousand people who are packed together like sardines, and who are all running in a particular direction. It makes no difference to one's trajectory of travel if one decides that one would like to move in the opposite direction, one will still be carried along with the crowd. It is only when a large enough number of the thousand people decide that they would like to move in the opposite direction that one's desire will translate into action. Analogously, only when the time is right will *the force to environmental sustainability* be powerful enough to challenge *the force to environmental destruction*. And the timing of this time will not be affected by the desires or the actions of particular individuals. Another way of putting this is to say that if an individual, or a family, or an entire town, suddenly acts one way rather than another (for example, the individual/s concerned might stop driving, flying, buying imported food, and so on) then this will make no difference to the way that the future trajectory of human cultural evolution will unfold.

We really need to face the reality of the situation we are in rather than put our hope in trivial and futile gestures which are ultimately insignificant. Recall that there are two categories of environmental problems (see *Chapter Seven*). When I talk of "trivial and futile gestures" I am referring to actions which are aimed at addressing the problem of global warming (the first category environmental problem). An action which is trivial and futile when it comes to this problem might not be so when it comes to the second category of environmental problems (all non-global warming environmental problems). For example, using reusable shopping bags is trivial and futile when it comes to addressing the problem of global warming, but it is not trivial or futile when it comes to other environmental problems, such as the problem of escalating waste in landfill sites.

The reality of the situation, in the face of the global warming problem, is that the human species needs to actively technologically regulate the temperature of the atmosphere of the Earth for the benefit of life on Earth. This is a positive outcome, not a measure of last resort which should be adopted out of despair. The sooner this is fully realised and accepted the better. When this time comes we can stop pursuing futile carbon-cutting policies and replace them with the required geoengineering policies.

Global Warming & the Anthropocentric and Ecocentric Attitudes

I think it is worth reflecting on some of the terminology which surrounds both the global warming debate and environmental issues more generally. When I became aware that the human species needs to actively regulate the temperature of the Earth's atmosphere, and that this is a good thing for life on Earth, I wasn't aware of much of the terminology that people have created when they debate such issues. For example, I hadn't heard of the terms 'geoengineering', 'anthropocentric', 'ecocentric' and 'the Anthropocene'. When I wrote my first book in 2010 – *Is the Human Species Special?: Why human-induced global warming could be in the interests of life* – I simply referred to 'the need to technologically regulate the Earth's atmospheric temperature'. I find most of the terms that have been created to be unhelpful. If the objective is to achieve a clear understanding of the situation we are in and how we need to respond to it through selecting appropriate environmental policy responses, then it might be best to use as little terminology as possible. Rather than talk of 'the Anthropocene', 'geoengineering', 'anthropocentricism' and 'ecocentricism', we could use simple language. If we are to use such terms, we need to be very clear about exactly what they mean, leaving little room for misinterpretation.

An unfortunate effect of the creation of such terms is that they provide a mechanism for causing division; once people associate themselves with a particular term, they typically become closed-minded to the realities inherent in alternative positions. For example, when one concludes that one must be 'ecocentric', then one will be naturally hostile to anyone who they believe to be 'anthropocentric'; for, as these terms have come to be typically used they are thought of as opposites; you are one or the other. So, the task of the 'ecocentric' (as they see it) is to persuade the enemy – the 'anthropocentric' – that they should desert their associates and switch their allegiance to a more enlightened camp. But the terms, as standardly defined, aren't actually even opposites! We would surely be a lot better off without all of this jargon. Here are standard definitions of ecocentricism and anthropocentricism:

> *Ecocentricism:* A philosophy or perspective that places intrinsic value on all living organisms and their natural environment, regardless of their perceived usefulness or importance to human beings.

> *Anthropocentricism:* A philosophy or perspective that sees human beings as the most important feature of the universe.

As I have already noted, these terms are typically used as opposites. I have been in debates with academic environmental philosophers who identify themselves as 'ecocentric'. On the basis of some of the things that I say they quickly conclude that I am 'enemy', an 'anthropocentric', and that I therefore cannot be a fellow 'ecocentric'. This causes them to say things such as *"You anthropocentric, you don't even care about any of the non-human life-forms on the Earth!"* This really is quite ridiculous, but you can see that the term 'anthropocentric' is hurled around by some people as a term of abuse.

This is ridiculous because anyone can see that the terms, as standardly defined above, are not actually opposites. One can believe that the human species is the most important feature of the universe, whilst also believing that all living organisms and their natural environment have intrinsic value, regardless of their perceived usefulness or importance to human beings. This is my belief. I am an anthropocentric and an ecocentric. Those who see an inevitable duality have deluded themselves. The terms are only opposites if one believes that humans have intrinsic value and non-human life-forms are devoid of intrinsic value. *In reality, all life-forms have intrinsic value, but the human species is the most valuable, most important, life-form on the Earth.*

It is one thing having philosophies and perspectives (ecocentricism and anthropocentricism), but when it comes to global warming and the environmental crisis, what is important are actions. When we look at collective human actions (at the global scale) are these actions in the interests of humans? Are these actions in the interests of non-human life on Earth? These are the important questions.

Most people seem to assume that humans are opposed to non-human life on Earth in such a way that the vast majority of actions which are in the interests of humans are not in the interests of non-human life. For example, if humans chop down part of the Amazon rainforest for agriculture, this is in the interests of humans, but it is not in the interests of non-human life. Such a view is narrow and simplistic. This is because it focuses on one particular event at one particular time, whilst the human relation to the non-human life-forms of the Earth needs to be seen in a collective global way which spans large swathes of time. If one focuses solely on a single leaf, one will be blind to the larger reality of the tree, and one will be utterly ignorant of the wonderful forest. Let us give up the leaves and gaze upon the forest.

>Anthropocentric actions = Collective humans actions (at the planetary level, over time) which are in the interests of the human species.

>Ecocentric actions = Collective human actions (at the planetary level, over time) which are in the interests of the totality that is life on Earth.

>Anthropocentric actions = Ecocentric actions.

This is the simple truth of the forest. In other words, the way that the human species has interacted with the planet is simultaneously in its own best interests and in the best interests of life on Earth.

George Monbiot on Atmospheric Carbon Dioxide Concentrations Reaching 400ppm

George Monbiot has been reflecting on the fact that atmospheric carbon dioxide concentrations have reached 400 parts per million for the first time in 800,000 years. Monbiot claims that:

> The only way forward now is back: to retrace our steps along this road and to seek to return atmospheric concentrations to around 350 parts per million, as the 350.org campaign demands. That requires, above all, that we leave the majority of the fossil fuels which have already been identified in the ground. There is not a government or an energy company which has yet agreed to do so.
>
> ('*Via Dolorosa*', 10 May 2013, http://www.monbiot.com/2013/05/10/via-dolorosa)

In effect, Monbiot is claiming that the only way forward is X, and then in his article he presents a good case that X is never going to happen. The conclusion is that climate breakdown (and the following extinction of the human species) is almost inevitable. So, Monbiot concludes:

> Without a widespread reform of campaign finance, lobbying and influence-peddling and the systematic corruption they promote, our chances of preventing climate breakdown are close to zero.
>
> So here we stand at a waystation along the road of idiocy, apparently determined only to complete our journey.

What is the cause of this tragic situation? Who is to blame? According to Monbiot the dire situation we face is due to the power of the fossil fuel companies, and the unsavoury character of those in high office:

> The problem is simply stated: the power of the fossil fuel companies is too great. Among those who seek and obtain high office are people characterised by a complete absence of empathy or scruples, who will take money or instructions from any corporation or billionaire who offers them, and then defend those interests against the current and future prospects of humanity.

It is hard to take Monbiot's assessment of the situation very seriously. I mean, let us start with the fact:

George Monbiot on Atmospheric Carbon Dioxide Concentrations

- *Atmospheric carbon dioxide concentrations on the Earth have reached 400ppm due to the combined effect of human actions over a prolonged period.*

There are obviously reasons why this state of affairs has come about; factors which have caused this situation. What are these factors? According to Monbiot, the causal factors are as follows:

- *The people in high office are characterised by a complete absence of empathy or scruples.*

I presume that you can easily see how unsatisfactory such an answer is. What are the actual causal factors which have caused the state of affairs to come about?

- *Since the first humans came into existence humans have been acting in accordance with the inner drives and feelings which are endowed to them by the universe. Because humans typically act in accordance with these drives/feelings human culture has evolved from hunter-gatherer to globalised technological society. When a planet becomes technological a side-effect of this is that the planet goes through a transitory stage of environmental disruption; this is because becoming technological requires the widespread appropriation and transformation of planetary resources. One of these transitory*

> *disruptions is a short-term increase in atmospheric carbon dioxide concentrations. Furthermore, when a planet becomes technological this is a sign of a healthy planet. In other words, if we were to detect a rise in atmospheric carbon dioxide concentrations on a distant planet, a rise similar to that which is currently occurring on the Earth, then we could safely conclude that this is a life-bearing planet where life is thriving. This is because the rise means that not only has life colonised the entire planet but it has also reached the stage of the birthing of technology. What a cause of joyous celebration!*

Monbiot is completely missing the bigger picture. He claims that we are determined to complete a journey along a "road of idiocy" because of a few individuals who are "characterised by a complete absence of empathy or scruples". The reality is that the planet is currently, thanks to the human presence, positively thriving. And the situation we are in cannot even remotely plausibly be attributed to a few people who have an absence of empathy or scruples; the situation we face is due to the drives/feeling states of billions of past and present humans.

The future is not a gloomy one, it is simply one in which the human species actively technologically regulates the temperature of the atmosphere for the benefit of life on Earth. There are a range of other benefits for life on Earth which also arise from the human-initiated technological birthing process, as we saw in *Chapter Eight*.

The Three Questions & the Philosophical Worldview

Human-induced global warming obviously has a particularly important place in my philosophical worldview. If one was to come across some of these articles without being aware that the discussion of global warming is part of an overarching philosophical worldview, and that the need for geoengineering the temperature of the atmosphere falls out of this worldview, then there is a good chance that one might be bemused or confused; one might even find the views expressed to be obviously wrong, or even incomprehensible.

Why is there a good chance that this would be one's response? This is likely because the dominant contemporary worldview concerning the relationship between the human species and the Earth, the view which dominates media/culture/politics, is very different to my worldview. Furthermore, the fact that the appropriate response to global warming, and the 'environmental crisis' in general, is wrapped up in a philosophical worldview, is barely even realised or discussed. In other words, the discussion of the appropriate response to global warming standardly takes place firmly within a dominating worldview, a worldview whose existence pervades the thought of the participants without them even realising that this is so. This dominant worldview is the 'Destroyer Worldview' which I outline in my book *Saviours or Destroyers: The relationship between the human species and the rest of life on Earth* (2012).

If you are unfamiliar with talk of 'worldviews' then a simple example should help. When the first pioneers realised that the Earth revolves around the Sun, the dominant worldview was that the Earth was the centre of the universe. This dominant worldview was so all pervading that the vast majority of people simply ridiculed the ideas of the first pioneers. The dominant worldview so pervaded the thought of the vast majority of people that they couldn't comprehend how that view could possibly be wrong; the pioneers were surely crazy people, thought the masses. The masses hadn't even entertained the possibility that their beliefs were simply a philosophical worldview; their beliefs seemed to be so self-evidently true that they weren't something which needed to be thought about or discussed. As with the dominant contemporary 'Destroyer Worldview', the 'Earth-centred Worldview' was so powerful that its existence wasn't even realised (it was taken to be a self-evident fact not a challengeable construction of the human mind) until it was initially challenged and ultimately shown to be utterly false.

So, with this in mind, here is the basic outline of my philosophical worldview. It includes the three questions which are of central importance to my view, an overview of the view, and some practical implications which fall out of the view. This information is reproduced from my page on Wikipedia (this is why it contains phrases such as 'this view is outlined by Cummins in his first book').

The Three Questions

- Does the human species have an important place on the planet?

- If the human species has an important place on the planet, how does this relate to the environmental crisis and human-induced global warming?

- How does all of this relate to the decisions and actions of individual humans?

Cummins claims that these three questions are deeply interrelated. The human species has an important place on the planet and this is deeply related to the environmental crisis and human-induced global warming. Furthermore, individual humans have been endowed with certain feelings/motivations by the universe; most humans act in accordance with these feelings/motivations and this ensures that human culture has an evolutionary trajectory towards human-induced global warming and planetary geoengineering.

The Philosophical Worldview

1. The Earth and the Sun, like all other parts of the Universe, are ageing entities.

2. The Universe is divided into two parts – life and non-life.

3. The entire universe is pervaded with states of feeling (panwhat-it-is-likeness).

4. Life is a good state of feeling for the universe to be in.

5. Life, and complex life in particular, requires certain conditions in order to survive.

6. When life arises it strives to stay in existence by spreading out over the planet it arises on and by regulating the temperature of that planet's atmosphere to keep it favourable for its continued existence (as described in Sir James Lovelock's *Gaia Theory*).

7. As the Sun's energy keeps on increasing the point will come when, in the absence of a technological species, the ability of life to regulate the temperature of the planet's atmosphere will cease.

8. In order to survive non-human-induced global warming life needs to evolve a technological species.

9. Becoming technological entails the opening of a division – part of the universe has to see itself as 'non-natural' and as opposed to 'nature'. Becoming technological also involves manipulating the 'natural' to the extent that an environmental crisis and technology-induced global warming are generated.

10. On the Earth the human species is that part of life which is technological.

11. The purpose of the human species is to be the saviour of life through developing and deploying the technology which regulates the temperature of the Earth's atmosphere.

12. Individual humans, just like all parts of the universe, naturally act/move in a way that maximises their state of feeling (towards 'pleasure' and away from 'pain'). The majority of humans acting in accordance with their feelings gives the trajectory to human culture which ensures that the human species fulfils its purpose.

13. The human species is special because it is the technological saviour of life on Earth.

14. The human species will fulfil its purpose because of concerns about, and/or the reality of, human-induced global warming.

15. This is the only reason that the human species is special ('raised up above' all of the non-human life-forms of the Earth). The human species isn't special solely because of rationality, consciousness, spirituality, language, culture, tool use, or any other 'unique' attribute.

This view is outlined by Cummins in his first book: *Is the Human Species Special?: Why human-induced global warming could be in the interests of life* (2010). He is also the founder of the paradigm of 'panwhat-it-is-likeness'. According to this view, mind and consciousness are very rare attributes in the universe, but states of what-it-is-likeness pervade the universe. The 'panwhat-it-is-likeness' view is developed in detail in the book: *An Evolutionary Perspective on the Relationship Between Humans and their Surroundings: Geoengineering, the Purpose of Life & the Nature of the Universe* (2012).

Panwhat-it-is-likeness is best thought of as different to panpsychism as the latter implies that psyche or mind or consciousness pervades the universe. In accordance with the Buddhist Theory of Atoms, Cummins contends that smells/tastes/feelings pervade the universe, and that only two senses evolve in humans – seeing and hearing.

Practical Implications

In *Saviours or Destroyers: The relationship between the human species and the rest of life on Earth* (2012) Cummins describes how we are living through a painful technological birthing process; life is bringing forth 'technological armour' to help ensure its future survival. He claims that there are two implications which follow from his philosophical worldview:

1   That if we realise that that the universe/life is 'inevitably' moving towards a better state through this current transitional stage of a painful technological birthing process then we can act differently. We cannot stop the process, but we can reduce the suffering it involves (for humans and for non-human life-forms). He claims that resources can be focused on geoengineering the temperature of the atmosphere, and resources that are currently being used trying to avoid this outcome can be more fruitfully deployed (these are wasted resources because not only can the geoengineering outcome not be avoided, it is actually a positive outcome). Resources can be redeployed to other pressing environmental and developmental issues which need to be dealt with at the surface of the Earth.

2   As parts of the 'feeling universe' (panwhat-it-is-likeness) we can seek to be more effective at living 'in tune' with our feelings and thereby effortlessly move to an optimal state of feeling – one that maximises our health and happiness. He claims that if one uses one's thought processes to 'disobey' one's feelings one will not be optimally happy. One should simply let the universe do its thing within one and thereby let the universe naturally move to the highest/best state of feeling.

Price's Response

In 2012, Peter Xavier Price, from the Sussex Centre for Intellectual History (University of Sussex), wrote a response to *Is the Human Species Special?: Why human-induced global warming could be in the interests of life* (2010). This is some of what Price had to say:

> Cummins' account of what he calls the 'trajectory of human evolution from hunter-gatherer to technological society' — indeed, the very thread upon which his whole argument is based—appears, in truth, to be little more than the eighteenth-century Scottish 'four-stages-theory', albeit in slightly modified form. Had Cummins acknowledged this interesting fact, he might even have reached the conclusion that we may now be entering (or already find ourselves in) a quinquennial, climatical phase of a potential 'five-stage theory', replete with its own conundrums and challenges.
>
> (*'Human Specialness': The Historical Dimension & the Historicisation of Humanity;* 2012, p. 41)

The Environmental Crisis & the Colonization of Space

I recently came across a copy of the *Resurgence & Ecologist* magazine which was published in late 2012 (September/October, No. 274). In this magazine there was one article which I think completely misses the point. In *'The Great Space Myth'* (pp. 54-5), John Naish attempts to convince the reader that, "the empty promise of space colonies only encourages the continuing ruin of the one planet we can inhabit". Naish claims that:

> such lofty ambition [to establish human space colonies] has a shadow side because it gives us permission to act as bad tenants of the planet we already live on. It encourages our species' habit of rapaciously destroying those ecosystems that support us only to abandon that mess and find new virgin territory to despoil.

It is the claim that the human activities (past and present) that have resulted in the 'environmental crisis' of modernity have been "encouraged" and "given permission by" the following belief that I find to be highly implausible; well, not highly implausible, just plain wrong:

> **The Belief:** If we change the Earth's biosphere to such an extent that it becomes uninhabitable then we can move to a space colony.

There is an 'environmental crisis' because this term is a concept created by humans to refer to a subset of human activities. The 'environmental crisis' has a cause or causes; this cause/s is the reason why humans have acted in the way that they have. What I find to be wrong is Naish's claim that one of the causes of the 'environmental crisis' is 'The Belief' (as detailed above). I believe that the causal roots of the 'environmental crisis' can be traced back to the origin of the human species, the origin of life on Earth, and even to the formation of the Solar System. However, many people simply trace the roots back to the Industrial Revolution; this was when the large-scale human modification of the Earth's biosphere was set in motion. I don't think that anyone believes that the pioneers of the Industrial Revolution were causally influenced in their activities by 'The Belief'. I don't believe that these pioneers thought something like: *we could do this Industrial Revolution thing and if it all goes pear-shaped we will simply move to a space colony!*

Similarly, I don't believe that *any* of the human actions that have resulted in the 'environmental crisis' have been causally influenced by 'The Belief'. The vast majority of humans, past and present, simply act largely in accordance with their inner feelings/motivations/drives; they act in a way which they believe will make them happy. For many humans alive today this means acting in a way which they believe to be sustainable/environmentally-friendly. For many humans alive today this means acting in a deeply unsustainable way (constant airplane flights, gas-guzzling cars, high all-round resource use). What seems wrong to me is the belief that those humans who are acting in an unsustainable way (individuals, corporations, governments) are causally influenced to act in this way by the belief that their actions are justifiable/permissible/acceptable because if everything goes wrong we can move to a space colony.

Clearly, if one wants to modify human activities across the planet in order to make them accord with what one personally believes to be more desirable (i.e. Naish, and a great many other people, have the belief that it is desirable for current human activities across the planet to be very different, to be more sustainable, etc.), then one needs to identify the actual causes of human activities. If one believes that human activities are caused by the belief that humans can escape to a space colony, then one is surely wrong.

Furthermore, from the perspective of my philosophical worldview, current human activities across the planet are just as they should be. *The force to environmental destruction* is in the ascendancy which is a sign that the Earth is bringing forth the technological protection that life needs in order to survive; *the force to environmental sustainability* is weak but growing, which means that when the technological protection is in place we can look forward to a long-term sustainable future on the planet. Looking into the distant future, space colonization seems to me to be 'inevitable'; human technology will enable both humans and non-human life-forms to live in non-Earth locations.

Technology and Stewardship

Two recent newspaper articles highlight the one-sided way that contemporary religious authorities have come to view environmental issues. The first article is *'Fracking risks God's creation, says Church'* by James Kirkup, which appeared in *The Daily Telegraph* (14 August 2013, p. 1). According to this article, a leaflet published by the Church of England Diocese of Blackburn states that "Fracking causes a range of environmental problems" and that believers should consider their Christian duty to act as "stewards of the Earth" in order to protect "God's glorious creation" from fracking. The leaflet also states that "succeeding generations" will suffer if "the Church remain[s] uninformed and silent" on the issue.

The second article is *'Reduce impact on climate or we disinvest, CofE warns companies'* by Sam Jones, which appeared in *The Guardian* (13 February 2014, p. 11). According to this article:

> The Church of England has said that it will, as a last resort, pull its investments from companies that fail to do enough to fight the "great demon" of climate change and ignore the church's theological, moral and social priorities.
>
> ... the Rev Canon Professor Richard Burridge, told the General Synod that... "Pointing the finger at the extractive industries avoids the fundamental problem which is our selfishness and our way of life, which has been fuelled by plentiful, cheap energy".

I will explain why this way of looking at environmental issues is one-sided. These assertions are based on the assumption that the Destroyer Worldview is true. This assumption gives rise to a particular interpretation of what it means for humans to be "stewards of the Earth". This assumption (that the Destroyer Worldview is true) is widespread in contemporary society, but one might have expected religious authorities to explore a different view, given that central to all religions is the importance of the human species, the specialness of the human species. That humans rightly have dominion over the resources of the Earth, that such dominion, such domination (a side-effect of which is the environmental impacts that it creates) is a good thing, is central to most religions. This is explicitly stated in the Bible and in many other religious texts. Such a view of rightful human domination leads to the Saviour Worldview, which is the opposite of the Destroyer Worldview.

Surely, one would think, religions such as the Church of England should be "informed" by their historical religious texts, rather than being "informed" by a particular contemporary social construction which is peddled by the scientists, environmentalists, politicians and activists of the day. Yet, when it comes to environmental issues it seems, based on the newspaper articles quoted above, that the religious texts have either been cast aside or reinterpreted in accordance with the dominant contemporary worldview (the Destroyer Worldview). In this case, the Church's desire to be "informed" seems to mean listening to what scientists and activists/protesters are saying, and giving this priority over what is said in their religious texts relating to the rightful dominion/domination of humanity.

If one initially approaches environmental issues not from the scientific/activist approach (the Destroyer Worldview), but from the theological/religious approach, then one can see things differently. However, the religious authorities seem happy to simply unquestioningly go along with the scientific/activist approach. In other words, they simply accept that *human environmental perturbations* and *human stewardship of the planet* are polar opposites rather than complementary.

If you look at many ancient religious texts for insight into the nature of the human/non-human relationship then the key phrase is 'human stewardship of the Earth'/'human dominion over the non-human life-forms of the Earth'. The scientific/activist interpretation of stewardship is that human technology is an evil, an encroachment on nature, an offence to (God's) creation/life. Contemporary religious authorities seem to have simply bought into this view. However, they need not, and they surely should not.

The religious texts offer a more compelling view of 'human stewardship'; a view according to which human technology is actually a 'gift from God'. Indeed, in this view, technology is the most precious part of God's creation, and humans, as the bringers forth of technology, are the most precious part of the Earth. God is the original creator, the bringer forth of the Universe; on Earth humans are the creators, the bringers forth of technology. So, humans have a special relationship with God and a special place on the planet.

If you look at the Bible, you will find that humans are special because of their technological abilities. When you read about Noah's Ark you will realise that the human technological ability to create, to bring forth The Ark, was a wondrous event which enabled humans to save the non-technological life-forms of the Earth. Technology is the saviour of life. I believe Noah's Ark to be a prophetic account. The human purpose on the Earth is to develop technology for the benefit of life on Earth. This is what 'human stewardship' entails.

What this means is that the development and deployment of technology is a fundamental part of human stewardship. Only certain technologies are required to fulfil God's purpose, but when the technological genie is released its development leads into all sorts of creations (nuclear power, fracking, airplanes, cars, submarines). Environmental problems are simply a deleterious side-effect of this purpose, this greater good, this bringing forth, this epoch of foretold human stewardship. In other words, environmental problems are a fundamental part of the human stewardship of the Earth; they are not antithetical to it.

I would urge religious authorities to consider the various interpretations of 'human stewardship' that exist; I would urge them to consider how the view that I have just outlined is supported by their religious texts. This would surely be much more fruitful than getting "informed" through the limited and skewed assumptions of scientists, politicians and environmental activists.

The Inevitability of Geoengineering

In several of the previous articles I have cited cutting-edge research and emerging opinions which provide evidence and support for my philosophical worldview. September 2013 saw a wave of further support for my view, notably from several articles in *The Guardian*. In this month the astronomer royal Lord Rees expressed the opinion that:

> If the effect [rising atmospheric CO_2/temperature] is strong, and the world consequently seems on a rapidly warming trajectory into dangerous territory, there may be a pressure for 'panic measures'... These would have to involve a 'Plan B' – being fatalistic about continuing dependence on fossil fuels, but combating its effects by some form of geoengineering.
>
> (The Guardian, *'Astronomer royal calls for 'Plan B' to prevent runaway climate change'*, Alok Jha, 11 September 2013)

In this month another relevant article appeared on *The Guardian* website entitled: *Why geoengineering suits Russia's carbon agenda*. In this article Professor Clive Hamilton states that:

There are some more reasonable Russian voices talking about geoengineering, including a handful of scientists modeling the impacts of sulphate aerosol spraying. However, they argue that geoengineering is inevitable because carbon emissions are growing by more than the IPCC's most pessimistic projections: "Therefore, humankind will be forced to apply geoengineering to counter the unwanted consequences of global warming."

(http://www.theguardian.com/environment/2013/sep/24/why-geoengineering-suits-russias-carbon-agenda)

I also came across another article in *The Guardian* by Professor Hamilton from 22 March 2013 (*'Why geoengineering has immediate appeal to China'*). In this article he cites the scientific evidence that the human perturbation of the atmosphere is currently accelerating rather than declining, levelling off, or declining:

> Yet neither China's efforts nor those of other countries over the next two or three decades are likely to do much to slow the warming of the globe, nor halt the climate disruption that will follow. Global emissions have not been declining or even slowing. In fact, global emissions are accelerating.

On the 28 September 2013 another article (*Why has geoengineering been legitimised by the IPCC?*) appeared on *The Guardian* website in response to the latest IPCC report. In this article Jack Stilgoe says he is "scared" by the mention of the word 'geoengineering' in the report:

> To include mention of geoengineering, and its supporting "evidence" in a statement of scientific consensus, no matter how layered with caveats, is extraordinary. If I were one of the imagined policymakers reading this summary, sitting in a country whose politicians were unwilling to dramatically cut greenhouse gas emissions (i.e. any country), I would have reached that paragraph and seen a chink of light just large enough to make me forget all the dark data about how screwed up the planet is. And that scares me.
>
> (http://www.theguardian.com/science/political-science/2013/sep/27/science-policy1)

One theme which underpins all of the opinions expressed in these various *Guardian* articles is the assumption that the technological regulation of the temperature of the atmosphere should be seen as a 'weapon of last resort'; it should be seen as something that we do only because we are incapable of sufficiently reducing our emissions of greenhouse gases. I have commented on this very widespread view of geoengineering as a 'weapon of last resort' in previous articles. All of the advocates of geoengineering that I have ever come across (except for me!) stress that they are only reluctantly advocating the measure as a regretful 'weapon of last resort'; in other words, if human greenhouse gas emissions could be magically slashed overnight then this would be preferable to geoengineering.

One of the most important conclusions that falls out of my philosophical worldview is that the technological regulation of the temperature of the Earth's atmosphere is a joyous event which should be actively and vigorously pursued. Such an activity is required for the continued existence and flourishing of the life that has arisen on the Earth. Such an activity is in the interests of life. If humans were not to carry out this activity then they would be condemning the Earth to a barren and lifeless existence.

I believe that we are likely to technologically regulate the temperature of the atmosphere in the widespread belief that such an action is a 'weapon of last resort' in response to irresponsible human activities; and that then, at a future date, we will come to realise that such an activity was actually a positive joyous event. We may even widely come to appreciate that the carrying out of such an activity was actually the reason that we came into existence as a species. However, it is at least possible that the 'collective awareness' of our place on the planet reaches such a level that the joyous and positive nature of geoengineering is realised before we carry out the activity. As I have noted before, such a realisation would have many benefits, including saving the enormous amount of money which is spent on futile schemes which attempt to avoid the need for geoengineering. Such money could be spent much more wisely on both geoengineering projects and other environmental and developmental projects. So, the recognition of geoengineering in the IPCC report is a very small step in the right direction.

The Conceptual Framing of Geoengineering

In the last article I considered the most recent wave of realisation concerning the need for geoengineering. Two weeks after I wrote this article a story appeared in *The New Scientist* magazine which outlines the specific types of geoengineering that can be deployed; the article also specifies the locations on the planet where each of these types can be implemented. Here is a taste of the article:

> THIS is how we will hold off disaster. To help us avoid dangerous climate change, we will need to create the largest industry in history: to suck greenhouse gases out of the air on a giant scale. For the first time, we can sketch out this future industry – known as geoengineering – and identify where it would operate.
>
> The bottom line is that CO2-suckers are essential, but we also need to ditch fossil fuels quickly. It's that or climate havoc.
>
> (The New Scientist, *'Terraforming Earth: Geoengineering megaplan starts now'*, Michael Marshall, 9 October 2013)

The Conceptual Framing of Geoengineering

I am glad to see that concrete plans are being made concerning how the technological regulation of the atmospheric temperature will be achieved. What we really need to see, in tandem with this, is a growing realisation that not only is this is an inevitable outcome, but that it is also a positive outcome.

The current widespread conceptual framing of geoengineering as a 'weapon of last resort' leads to inevitable resistance to the phenomenon. There even seem to be a surprisingly large number of people who are so opposed to geoengineering that they would rather see a massive jump in the atmospheric temperature of the Earth which instantaneously wipes out complex life on Earth, than they would see geoengineering deployed to stabilise the atmospheric temperature for the benefit of life on Earth. Such a view is irrational and it is potentially harmful; its existence is one reason why I believe that it is helpful to move to a different conceptual framing of the phenomenon.

If we move to a conceptual framing which is more closely in accordance with reality then we can reap the benefits which this entails. We can adopt a philosophical worldview which entails that the life that has arisen on the Earth is currently giving birth to the technological armour – via the human species – that will enable it to survive and thrive way into the distant future. It is this philosophical worldview that I have been outlining throughout this book.

The Technological Healers of the Earth

In the July/August 2013 edition of the *Resurgence & Ecologist* magazine Charles Eisenstein outlines his view concerning the relationship between environmental problems, technology and the healing of the planet (*Latent Healing,* pp. 36-8). Eisenstein has many interesting things to say; however, the central premise of his view is fundamentally flawed. In this article my aim is two-fold. Firstly, I will explain why the central premise of Eisenstein's view is grounded in a false dichotomy. Secondly, I will comment on the positive aspects of Eisenstein's view and outline how it differs to my philosophical worldview.

So, let us start by identifying the false dichotomy that exists at the heart of Eisenstein's view. Eisenstein states that:

> We can assume that by now the environmentally conscious person has seen through the delusion of applying technology to remedy the problems that have been caused by previous technology. (p. 36)

> The technological fix addresses the symptom while ignoring the illness, because it cannot see an integral entity that can become ill. I don't want to gloss over the profundity of the paradigm shift we are accepting if we are to see Nature as intelligent and purposive. (p. 37)

Furthermore, Eisenstein claims that anyone who believes in a "technological fix" to environmental problems is constrained by a "mythology" which causes them to exist in a "disconnected state of being that is blind to the indwelling purpose and intelligence of Nature". (p. 37)

Whereas, according to Eisenstein, he isn't stuck in such a "mythology". Apparently he has "seen through this delusion" and is therefore an "environmentally conscious person" who can see that we live in "an inherently purposeful universe". He claims that: "The technological fix is based on linear thinking. The alternative is to develop sensitivity to the emergent order and intelligence that wants to unfold, so that we might bow into its service." (p.38)

So, in short, Eisenstein attempts to persuade us that these two things are sharply antithetical:

1   A purposeful intelligent universe

2   A technological fix to environmental problems

This belief is completely and utterly wrong; he has simply set up a false dichotomy based on *his own mythology*. The question of whether there is a technological fix to environmental problems *is* very plausibly influenced by the question of whether the universe is purposeful and intelligent. However, it is just as plausible to believe that it is precisely *because* the universe is purposeful and intelligent that an environmental 'technological fix' is required. Eisenstein doesn't even appear to realise that this is a possibility. So, it will surely be helpful if I relate what Eisenstein has to say to the philosophical worldview which I have been developing over the past decade. There are some interesting commonalities between Eisenstein's view and my philosophical worldview. However, it seems obvious to me that my view is much deeper and more comprehensive. In other words, Eisenstein is on the right track but he hasn't had the deeper insight which would have enabled him to put the pieces of the jigsaw together to form a more complete cosmic picture (he is missing several pieces).

Both Eisenstein and I support the idea of a purposeful intelligent universe, the unfolding of which includes the bringing forth of the human species. The main difference in our views is that I have seen how technology, and in particular the application of human technology in the environmental arena, is a fundamental part of a purposeful intelligent unfolding universe. Indeed, it is obvious to me that our purpose as a species is to utilise technology in order to regulate the temperature of the atmosphere; this outcome being for the benefit not just of humans, but for life on Earth. Eisenstein hasn't come to appreciate this, but unlike most people, he is pondering closely related questions, such as (p. 38):

> What is the purpose of technology on a healed planet?

> What is the purpose of this unique species [humans] to which Gaia has given birth?

The first question Eisenstein poses actually makes no sense to me because it is obvious to me that the purpose of technology *is* to heal the planet; it would have no purpose on a healed planet. Of course, technology would have uses on a healed planet, but uses and cosmic purpose are two very different things.

A central component of my philosophical worldview, a component that is lacking in Eisenstein's view, is the realisation that the planet was ill/required healing *before* it gave birth to humans. In other words, it is *not* the case that technology created a problem (and that even more technology cannot provide the solution to this problem). Einstein's view is seemingly grounded in the belief that if technology creates a problem then it cannot simultaneously solve that problem. This seems highly dubious to me as a blanket view, but even if it were true then it doesn't apply to what we are talking about here. For, the need for technological regulation of the temperature of the Earth's atmosphere is ultimately a non-technological problem; a non-technological problem to which technology is the solution. Of course, this simplifies what is a complex situation, because the deployment of human technology has exacerbated the pre-existing problem/illness. The key point to realise is that humans, and human technology, are not the cause of planetary illness. Exacerbating a pre-existing illness is a very different kettle of fish from creating an illness. Indeed, exacerbating a pre-existing illness can be a necessity if that illness is to be cured.

I presume that you can see what I am saying here. For the sake of argument, let us grant Eisenstein his claim that technology cannot be the solution to a problem created by technology. This means that all of the environmental problems which have been caused by human technology will have no technological solution. Nevertheless, if there is an environmental problem which has its roots in a non-human pre-human cause then, even if human technology exacerbates that problem, it can still be the solution to the problem. Indeed, the exacerbation can be a sign that the cure is imminent. In the rest of this article I am hoping to get you to see that global warming is such a problem, and that human technology is the solution.

At the heart of Gaia Theory is a vision of the Earth as a self-regulating but ageing whole. Another way of putting this is to say that the Earth attempts to maintain the conditions suitable for life despite the ageing of the Earth-Solar System. The Earth needs to maintain a Global Mean Surface Temperature (GMST) which is neither too cold nor too hot if complex life (plants and animals) is to exist; this 'habitable' temperature range is between 10°C and 20°C. The Earth has managed to achieve this GMST throughout the history of life on Earth despite an increase in incoming solar radiation of 25% since life arose. As Sir James Lovelock puts it:

> We may at first think that there is nothing particularly odd about this picture of a stable climate over the past three and a half eons [3,500 million years]… Yet it is odd, and for this reason: our sun, being a typical star, has evolved according to a standard and well established pattern. A consequence of this is that during the three and a half aeons of life's existence on the Earth, the sun's output of energy will have increased by twenty-five per cent.
>
> (*Gaia: A New Look at Life on Earth*, OUP, 2000, p. 18)

The Technological Healers of the Earth

As the Earth-Solar System ages a forever increasing amount of solar radiation is sent from the Sun to the Earth; this is obviously a force for global atmospheric warming. The Earth-Solar System is now at the age where the Earth is struggling to maintain its atmospheric temperature within a range in which complex life can survive. The increasing solar radiation is putting immense upwards pressure on the Earth's atmospheric temperature which could lead to an increase to a level which is too hot for complex life to survive. This struggle is a sign of planetary illness. Lovelock has realised this:

> The brief interglacials, like now, are, I think, examples of temporary failures of ice-age regulation.
>
> (*The Revenge of Gaia*, Penguin Books Ltd, 2006, p. 45)

> [Gaia] is old and has not very long to live. As the sun grows ever hotter it will, in Gaia's terms, soon become too hot for animals and plants and many of the microbial forms of life.
>
> (*The Revenge of Gaia*, Penguin Books Ltd., 2006, p. 46)

So, the transitions between ice ages and interglacial periods are an indication of planetary illness which is caused by the increasing output of the Sun. We are now approaching the time when the Earth requires technological regulation of the temperature of its atmosphere. The Sun will continue to send more and more solar radiation to the Earth. The deployment of technology is the only way that the temperature of the atmosphere can be kept suitably low for complex life to survive. Such a use of technology is the only way that the Earth can be healed.

In other words, the Earth has brought forth a technological species precisely at the moment in its evolution when it requires technological regulation of its atmosphere. This is surely a sign that we live in a purposeful unfolding universe; a universe which contains humans as the healers of the Earth. What a wonderful vision! The development of the healing technology is itself a painful process, it entails the bringing forth of a range of environmental problems and also the transitory human exacerbation of the illness for which human technology is the cure. However, this era of human separation, of suffering, of environmental problems, of technological development, is a transitory era which results in the healing of the Earth and ultimately the healing of humanity too.

So, in contradiction to what Eisenstein claims, a purposeful intelligent universe can be one in which there is a need for technological geoengineering of the atmospheric temperature (Eisenstein rejects this and claims that it "will likely cause horrific unanticipated consequences", p. 36). Whilst Eisenstein is wrong when he denies that a purposeful intelligent universe can entail the need for geoengineering, he is surely right to assert that many of the environmental problems that we face are caused by technology and have no technological solutions.

The vision I have presented of humans as the technological healers of the Earth clearly does not deny that humans have caused environmental problems which need to be addressed. Technology has caused many environmental problems and I believe that there is a space for both technological and non-technological solutions to these problems. The human role as technological healers of the Earth is a specific role: regulating the temperature of the atmosphere so that life can survive and thrive despite the increasing output of the Sun. As the technological healer of the Earth the human species is the saviour of life on Earth.

The Concept of 'Future Generations'

If you are familiar with ethical debates concerning the environmental crisis then you will be aware of the way that the concept of 'future generations' is regularly called upon in such debates. The concept is typically used as follows:

A) Recent and current human activities across the planet are of such a magnitude that the planetary conditions that exist when future generations come into existence seem likely to be very different to the conditions that exist today.

B) These future conditions will be far less hospitable and pleasurable than those that exist today.

C) Humans that currently exist owe it to future generations to change the way that they act, so that they can 'hand over' a planet which is just as hospitable and pleasurable as it is now.

The term *future generations* is used with varying scope. In its limited use it refers only to humans, so we are here talking about the impacts on the grandchildren of our children, the grandchildren of the grandchildren of our children, and other humans that might potentially exist hundreds and thousands of years in the future. In its broader use the term also refers to non-human planetary life-forms. In this extension of the term it is the offspring of the offspring of the offspring of currently existing whales, cats, bees, seagulls, tigers and elephants, and so on, which is being considered. These potentially existing future life-forms are *future generations.*

Whatever the scope of the term, the concept of *future generations* is used in different scenarios concerning the future potential state of the Earth. There are two standard scenarios in which the concept is utilised:

- Recent and current human activities are so destructive that the planet that future generations inherit will be a far less pleasant place to live than the planet that we currently live on. We currently have an ample supply of food and water (even if it is unequally distributed), we have beautiful natural wonders to gaze upon such as majestic forests, and we have an abundance of wonderful non-human life-forms to interact with. However, we are currently using so much of the Earth's resources, and are transforming the Earth's habitats to such an extent, that future generations seem likely to inhabit a comparatively barren world. Their resources will be far fewer, their living standards lower, the natural beauties upon which they can gaze their eyes will be vastly diminished. And much the same can be said for future generations of non-human life-forms. Surely, the argument goes, we owe it to future generations to ensure that this state of affairs does not come about.

- Recent and current human activities are so destructive that if we don't have a vast and near-immediate change in our interaction with the planet there will be no future generations. Hundreds of years from now there will be no humans on the Earth; there will also be no elephants, no tigers, no cats, no dogs, no bees and no whales. A future generation-less planet awaits us. Surely, the argument goes, we owe it to the life-forms that could exist in the future, to act in a way which ensures that the conditions exist which enable them to be born.

There are numerous intermediate positions which combine elements of these two scenarios. For example, one could believe that unless human activities significantly change that the future human population size will drop by ninety per cent and the remaining ten per cent will live on a fairly barren planet which has much lower living standards than we enjoy today. So, the term *future generations* is deployed in environmental discussions to refer to two future possibilities:

1 The possible future *existence or nonexistence* of life-forms which currently do not exist.

2 The *quality of life* of the life-forms that exist in the future.

The question I think we should consider is: Should we really be concerned about *future generations?* In other words, is the concept of any value? Is it useful to imagine the life-forms that might possibly exist in the future and try and gauge what their quality of life is likely to be? I am not convinced that the concept is particularly useful. For one thing, we have a long way to go when it comes to the welfare of the life-forms (both human and non-human) that actually exist; we could fruitfully concentrate on these life-forms rather than speculating about the quality of life of non-existent entities, imagined entities that might possibly exist hundreds of years in the future. I do definitely believe that we should act in a way that ensures that life on Earth can survive and thrive in the future. However, talking of 'life on Earth' seems to me to be sufficient; there doesn't seem to be any need to imagine *future generations* and whether we have a duty to them.

Those who create arguments around the concept of *future generations* seem to believe that we have a duty to *particular individual* non-existent life-forms, the grandchild of the grandchild of their grandchild, the grandchild of the grandchild of the grandchild of their pet tortoise, and so on. It is surely more helpful to think along the following lines. Firstly, let us have respect for every human, and for every life-form, that actually exists; let us not speculate about which possible life-forms might or might not exist in the future. Secondly, let us consider what actions we need to take to ensure that the Earth can continue to sustain life, complex and interesting life, into the distant future. If we ensure that the future Earth is in a healthy state, a state which enables life to survive and thrive, then we will have done our job. The particular life-forms that exist in the future will be more than capable of looking after themselves.

The Philosophy of Global Warming

Such a consideration leads us to the conclusion that the most essential thing that we need to do is to technologically regulate the temperature of the Earth's atmosphere. Such an activity forms the bedrock which will enable life on Earth to survive and thrive in the future.

***

You will probably recall that there is a little more on the issue of *future generations* in the Dialogue section of the book (p. 213), where our objector asked: *Do you think that we have a responsibility to 'future generations'?*

Is Fracking Good or Bad?

Fracking has emerged as a new area of environmental confrontation, particularly in the UK and the US. On one side of the confrontation are environmental activists who are passionately attempting to stop fracking; on the other side are the businesses engaged in fracking and the government, which is subsidising their activities. In this article I outline how the fracking confrontation relates to my philosophical worldview.

In this book I have often referred to two forces which are driving the evolution of human culture – *the force to environmental destruction* and *the force to environmental sustainability*. I initially outlined the nature of these forces in *Chapter Seven*. These two forces are represented by the two sides of the fracking confrontation:

- *The force to environmental destruction* – The businesses engaged in fracking, the government which is subsidising fracking, the millions of people who like to use lots of energy and like it to be as cheap as possible, and the transnational flows of technology and resources which cause governments to seek to be competitive on the 'world stage' (which requires utilising the latest technologies and having cheap energy costs for business).

- *The force to environmental sustainability* – Those people (individuals, groups, charities) who passionately believe that fracking is a bad thing because of potential environmentally deleterious consequences.

The force to environmental destruction is currently the dominant force in human culture; more than this, it is the underlying force which has driven the entire evolutionary progression of planetary life from simple beginnings billions of years ago to globalised technological society. It is an exceptionally powerful force. In contrast, *the force to environmental sustainability* is currently in its infancy; it is puny in comparison to the destructive force, it is weak but it will gradually grow in strength over time.

I should stress that these forces are just the workings of the universe as it gradually unfolds; there is no 'goodness' or 'badness' involved in the forces. *The force to environmental destruction* is not 'bad' simply because it involves the word 'destruction'. I could have used another phrase to refer to this same force; I could have called exactly the same force *the force to planetary ecstasy,* and this would be an equally appropriate name.

What does this mean when it comes to fracking? Well, the powerful force will be victorious. Fracking is already a major energy source in the US and it will become so in the UK. However, in some sense, talk of being 'victorious' misses the point. The environmental activists who are passionately opposing fracking should not be thought of as wasting their time. These activists, and *the force to environmental sustainability* of which they are a part, are a crucial part of the unfolding planet as it moves towards a more harmonious future. In the future, the Earth will be in a state where the two opposing forces (destruction/ecstasy and sustainability) are in a state of perfect balance (or to put it another way, the *singular force* will have fully matured). This state will only come about because of the growing strength of *the force to environmental sustainability*. In other words, despite losing the fracking confrontation, those who are passionate about the environment should continue to channel their energy into similar activities.

What I am saying here needs to be seen in the context of my wider philosophy. Environmental activists typically take a narrow view which focuses wholly on the effects of a singular activity. For example, a typical activist view would be: *fracking is bad because it involves risks such as contamination of the water supply, air pollution, and geological destabilisation; it will also contribute to global warming. We can get our energy from safe and renewable resources; we have no need for fracking.* My philosophy gives a wider perspective within which the human presence on the planet is a positive joyous one; this is because through the human species life on Earth is currently evolving itself a set of 'technological armour' which will ensure its future survival. Furthermore, our current use of fossil fuels is playing a major part in us fulfilling this objective of bringing forth the 'technological armour'.

If one takes a narrow view then one can make a case that fracking is 'bad'. However, if one takes the wider view then, if anything, fracking is 'good'. In other words, the fact that the Earth has reached the stage in its evolutionary progression where life is using and deploying fracking technologies means that life is positively thriving. So, in one sense fracking is neither good nor bad (see paragraph six of this article), and in another sense it is both good (a sign that the Earth is thriving) and bad (some local deleterious impacts).

There are dangers with all technologies – cars, guns, knives, helicopters, airplanes, trains, food processors, wind turbines (they kill lots of birds), drilling platforms and nuclear power stations. There have obviously been some very severe environmental impacts resulting from technology in the past. Given this, it is not surprising that people protest against new technologies, whether this is genetically modified food, nuclear power or fracking. It is also not surprising that many people often believe that the dangers of new technologies are more extreme than they actually are; a fear of the unknown can be a healthy thing.

The fracking debate is symptomatic of the wider relationship between the human species and the planet. The human species is that part of life which has become technological; it is the bringer forth of technology, the saviour of life. Having this pivotal place in the evolutionary unfolding of the planet is not necessarily a desirable one. Humans can suffer terribly as a consequence of being the technological animal; they suffer mentally (*Why I am here? How can I be happy? Why did my friend have to die at such a young age?*), and they suffer because of the dangers of technology itself. In the UK the fracking industry has to comply with high safety standards in order to prevent harmful environmental impacts. Despite this it is certain that fracking will be the source of suffering for some humans and non-humans (even if this is just due to the machinery being an eyesore and/or noise pollution); this is one of the reasons why people feel that they need to take part in fracking protests. This suffering is a local phenomenon, it affects individuals and local communities; at the national level there is no suffering but there are great benefits. This microcosmic example is analogous to the larger reality of the human presence on the planet; humans suffer in order to help the Earth move to a greater state of ecstasy.

Extreme Weather Events & Global Warming

I am writing this article in February 2014 in the midst of what many people are calling 'extreme' weather events in the UK. It has certainly been very stormy, with the highest amount of rainfall in the period since records began. Strong winds and abnormally high rainfall has led to coastal flooding and damage, and to numerous rivers bursting their banks. Many homes in the Somerset Levels have been flooded for over a month and there is currently no railway connection from Devon and Cornwall to the rest of the UK due to extensive storm damage to the railway infrastructure.

You won't be surprised to learn that many people are wondering whether there is a link between these 'extreme' weather events and human activities on the planet. To be a little more exact, the issue that keeps popping up in the media is whether there is a link between these 'extreme' weather events and human-induced global warming. However, the issue of real importance is slightly different to this. The issue of real importance is surely whether there is a link between these 'extreme' weather events and global warming (the aggregate of non-human induced global warming and human-induced global warming). If there is a causal link between these 'extreme' weather events and global warming, then this means that the immediate technological regulation of the atmospheric temperature would have immediate benefits; it would enable global warming to be stopped in its tracks and would thereby stop an escalation of 'extreme' weather events in the immediate future.

The mainstream view seems to be that *these 'extreme' weather events are probably caused by, or are at least made more severe by, global warming, but that this cannot be known for sure.* It certainly seems to be a fact that a warmer atmosphere will cause a change in the climate in various parts of the Earth, and that this change will include weather events which people consider to be 'extreme' compared to what came before. However, when it comes to the current weather events in the UK, one cannot conclude with certainty that the cause of these events is global warming. So, we are left with the mainstream view that the causal link is a 'probable' one.

Saying that something is 'probable', as is widely done with regards to this possible causal link, isn't particularly satisfactory. This could mean that the degree of certainty in the causal link is 50.1 per cent, or it could mean that the degree of certainty in the causal link is 99.9 per cent. Can we make any progress in thinking about the nature of this causal link?

The first thing to consider is timescales. People inevitably have a very short-term memory, because in the bigger scheme of things people don't live very long. The climate in the UK, and the associated nature of the weather events ('extreme' or 'mild') in particular locations, has varied immensely in the past. Over medium to long timescales big changes in climate are normal; such change is inevitable; such change is to be expected. When 'extreme' weather events occur they are typically labelled as such because people cannot recall many, if any, similar events in the handful of decades that they have been alive. Our cumulative weather records themselves only go back a few hundred years. Considerations such as these seem to lend weight to the idea that the current weather events which have been labelled 'extreme' are actually normal weather events; they are not caused by global warming.

However, this conclusion doesn't immediately follow. After all, global warming has been a phenomenon affecting the climate of the Earth since the Earth was formed. At a broad scale, the entire history of the Earth can be seen from the perspective of the interplay between non-human-induced global warming (the increasing output of the Sun) and responses to this phenomenon made on the Earth (responses which result in the homeostatic regulation of the atmospheric temperature). This interplay has inevitably been one of the main factors changing the climate (thereby generating 'extreme' weather) in particular parts of the Earth over decades, hundreds of years, and thousands of years. So, we cannot think of any 'extreme' weather event as being wholly divorced from global warming. Furthermore, this interplay has reached the stage in which the Earth's (non-technological) homeostatic regulatory capacity is weakening; given this current state of weakness, this current difficulty in 'smooth'/'easy' regulation of the atmospheric temperature, an increase in climate variability can be expected. In other words, the current 'extreme' weather events can be causally linked to non-human-induced global warming, not just to global warming.

The second thing to consider is that we are currently living through the epoch of technological birthing. As we have already explored, this entails the coming of a point of realisation for the human species:

- *The realisation of the extent of the perturbations that the human species has made to the Earth.*

This point is followed by three further realisations:

- *The realisation that the human species needs to deploy technology to regulate the atmospheric temperature of the Earth.*

- *The realisation that this is a wonderful thing for the totality that is life on Earth.*

- *The realisation that such regulation is the purpose of the human species.*

There is a gap between the initial realisation – the realisation of the extent of the perturbations – and the three later realisations. We are currently in this gap. Within this gap there is confusion about our place on the planet and our relationship to non-human life-forms. Within this gap there are also forces in play which seek to stimulate our progression to the later realisations. One of these forces is an increased number and severity of 'extreme' weather events. There are two factors of importance here:

1. The increase in the number and the severity of 'extreme' weather events caused by the reality of the perturbations caused to the biogeochemical cycles of the Earth by human activities in tandem with non-human-induced global warming.

2. Human concern that particular 'extreme' weather events are caused by human activities.

Either, or both, of these factors can stimulate our progression to the later realisations. The second of these factors is in play in the UK at the moment. The 'extreme' weather currently affecting the UK could have no causal link to global warming or human activities, yet this isn't important, it is the concern itself which can be a catalyst to the later realisations. In other words, the concern might or might not be reflected in reality and the reality isn't important; in this case being a catalyst is more important than the truth.

Let us now consider the first of these factors. One thing seems to be certain: If the human species is 'slow on the uptake', if it does not move speedily to the later realisations, if it ignores the concerns, if it doesn't push ahead with the technological regulation of the atmospheric temperature, then the first of these two factors will become more severe and pronounced until we get the message. In other words, the longer we delay technologically regulating the atmospheric temperature, the more 'extreme' weather events we will be letting occur; we will effectively be 'inviting' them to occur. So, whilst there is no certainty concerning the cause of the current 'extreme' weather events in the UK, there is certainty that global warming will cause a plethora of such events in the future if we are not successfully technologically regulating the Earth's atmospheric temperature.

How Much of Man is Natural?

This article contains an essay which I wrote in 2008. In that year the *Spinoza-Gesellschaft* organised an international prize essay competition which invited people to write an essay which addressed the question: 'How much of man is natural?' I submitted the following essay to the contest and it won the €1000 first prize (the essay which follows is slightly different to the winning essay). In my writing I usually use the terms 'the human species' or 'human' rather than 'man'; however, in accordance with the title of the competition, this essay uses the word 'man'. The word 'man' is used to refer to 'human' or 'the human species'.

***

The definition of natural is 'present in or produced by nature'. Is it not obvious to anyone who thinks about the question of 'how much of man is natural' that man has been *produced by* nature, and that every fibre of his being and existence is *present in* nature? Surely, a more appropriate question would be: "How could man possibly doubt that he is completely natural?"

In a trivial sense man, as a creator of words, can create a word such as 'natural', and define it in such a way that it excludes man. The word 'natural' can be opposed to *either* 'artificial' *or* 'supernatural'. However, interestingly, the word 'artificial' is defined as 'made by humans; produced rather than natural'. This definition does not refer to man *himself;* rather the word 'artificial' is itself an arbitrary construct, a word of use in human communication because it enables the *productive activities of man* to be referred to. There is no notion in the word 'artificial' that man himself is not natural. Furthermore, there is no implication that in the world itself there is a fundamental division in nature that the word 'artificial' refers to. It is simply of use to man to have a word that labels the results of his productive activities.

The notion of 'human production' is actually a deeply problematic one. If one gives the matter no real thought then the distinction between the 'natural' and the 'artificial' seems to be obvious, but reflection reveals otherwise. It is obviously the case that before the human species evolved nothing was artificial. It also seems obvious that objects such as the Sun and the planet Jupiter are not artificial. But when we focus on the Earth, then it is hard to identify anything that is truly natural. Let us consider a tree that is growing in a rainforest, and a tree that is produced in a human factory. One would be tempted to call the former 'natural' and the latter a 'produced artefact'. However, when one learns that the rainforest tree has the particular attributes that it has because humans soaked the surrounding ground with nutrients and breathed additional carbon dioxide into the vicinity of the tree, then one might have to concede that this tree is also an artefact.

Similarly, a human-constructed wigwam-shaped structure composed of branches would be a 'produced artefact', but a single branch lying on the ground under a tree would be considered 'natural'. Even if a human steps on the branch and breaks it the branch would still be considered 'natural'. But there is no difference in kind between modifying a branch by stepping on it, and moving several branches into a wigwam structure. At a larger scale, human activities have modified the climate and atmosphere of the entire planet thereby making the concept of the 'natural', as opposed to the 'artificial', largely redundant when it comes to the biosphere of the Earth. So, there is no meaningful distinction *in reality itself* between the 'artificial' and the 'natural'.

The word 'supernatural' is defined as: 'of or relating to existence outside the natural world'. It has to be questionable whether this word has any meaning whatsoever – surely all that exists is the natural world – *nothing* exists outside it. The word 'supernatural' is also used to refer to 'a power that seems to violate or go beyond natural forces', and also 'of or relating to the miraculous'. These descriptions are instructive because they imply that man uses the word 'supernatural' to refer to those parts of the universe around him that he cannot comprehend. So, in the past a total eclipse of the Sun would have been referred to as a supernatural event, a miracle. But now such an event is simply considered to be a natural occurrence. This means that the word 'supernatural' delineates man's *understanding* of how the universe works from how it *actually* works; it doesn't imply that in reality parts of the universe are not natural. A complete description of the universe would include solely 'natural' phenomena; there wouldn't be any 'supernatural' phenomena.

So, the word 'artificial' simply refers to the productive activities of man in the world, and the word 'supernatural' can be thought of as delineating the limits of man's understanding of the world. These terms are perfectly compatible with the belief that man is wholly natural, and the belief that all that exists is the natural world. Indeed, if one denies this and asserts that the world, and the Earth, contains natural parts and non-natural parts then one faces a seemingly inextricable problem. One has to try to untangle the complex and intricate way in which the activities of man have modified man's surroundings in order to separate what is natural from what is not natural. If one attempted to do this then one would surely conclude that the entire biosphere of the Earth is *not* natural. It is surely more sensible to conclude that the whole of the Earth is natural, and that *the word 'artificial' refers to things that, whilst produced by man, are still natural.* Alternatively, if one really believes that *the world itself* is split into the natural and the supernatural, then one has to specify exactly what these supernatural entities are. History suggests that there are no such entities because as understanding increases the 'supernatural' gets reclassified as natural.

Nevertheless, the very fact that man can doubt whether he is natural is clearly of interest. To understand the existence of this doubt we need to deconstruct the word 'natural'. Natural has been defined as: 'present in or produced by nature'. The word 'nature' itself has not yet been defined, but is contemporarily defined as: 'the material world and its phenomena'. This definition of nature sheds light on why man could possibly doubt that he is completely natural. It is the contemporary *conception* of nature as the 'material world' that causes man to doubt whether he is 'natural' because man himself doesn't seem to be composed of 'mere matter'.

In the face of this doubt there are two 'naturalistic' options. Firstly, one could hold that the term 'mere matter' is an accurate description of much of the world, but that man is not made of 'mere matter', and therefore that 'mere matter' has the ability to *produce* entities such as man which are not made of 'mere matter'. Secondly, one could hold that the term 'mere matter' is vacuous – there are no parts of the world that fit the term. In other words, one can hold that man's knowledge of 'nature' is still at a primitive level and that the label 'material world' is deeply misleading; it could be the case that all of nature has states analogous to those in man. Both of these options are 'naturalistic' options; both entail that man *is* wholly natural. However, if one's beliefs put one into the first group then it is much more likely that one will consider oneself to *not* be natural.

Most of the above analysis relies on the definition of words – definitions that can help explain why man might not consider himself to be natural, *given* the definition of natural. However, there is a much deeper issue as the definitions themselves are clearly expressions of man's pre-existing conceptual framework. The real issue is why man's conceptual framework in the current epoch is such that he defines nature as the 'material world and its phenomena'. Why do most contemporary humans consider themselves to be opposed to the surrounding world in some fundamental way? Could it be that it is a fundamental characteristic of what it is to be human to conceive of oneself as opposed to the surrounding world?

There are clearly two distinct issues. Firstly, given that the word natural means 'present in or produced by the material world and its phenomena', what could it possibly mean for man to be 'not natural'? Secondly, if man is wholly natural why does he doubt his 'naturalness'? Why does he consider himself to be opposed to the surrounding world? These two questions are obviously closely interrelated because it is the belief in an opposition between man and the surrounding world which leads to the conceptualizing of that world as 'mere matter'. If the belief in an opposition didn't exist then the natural world itself would, no doubt, not be conceptualized as 'mere matter'. Rather, it would be conceptualized in such a way that the attributes of man and world are tightly coupled.

Man's conception of the natural world

It goes without saying that man's conception of the natural world has varied immensely through time, and that it also varies between different cultures in the contemporary epoch. It is the dominant contemporary 'western view' of the natural world which I will focus on here. This view establishes a particularly sharp division between man and world, a division that is so deep that man's naturalness can be seriously doubted. According to the 'western view' the natural world is a clockwork mechanism the activities of which have been increasingly accurately predicted through science. The operations of the non-living world, and much of the living world, are conceived of as thoroughly deterministic and are wholly devoid of qualitative feeling, intentionality and awareness. In contrast, man has 'free will' to act in ways which cannot be predicted by science, and has qualitative feeling, awareness and intentionality.

Of course, there are those who live in the 'west' who don't subscribe to the 'western view'. Some people believe that the entire natural world has intentionality, or qualitative feeling, or awareness, and some people believe that quantum physics reveals that the entire natural world has freedom. Others argue that man himself doesn't have free will – it is simply an illusion. What is one to make of the claims of these people who oppose the 'western view'? The issues raised are very deep any many seem to be unanswerable. Who could say whether states of aboutness/intentionality exist when atoms interact to form molecules? Who could say whether these interactions themselves entail qualitative feeling? Who could say whether there is any kind of awareness present in these interactions? Who could say whether these interactions actually result from free will? And who could possibly know if every thought that they have ever had is determined, and in theory predictable before they had it?

I take it that no-one has adequate answers to these questions. Therefore, the question needs to be asked as to why the 'western view' always sides with the 'oppositions' – the answers to the questions which lead to an opposition between man and world. This 'siding' clearly says nothing about the natural world *itself* – all it reveals is the way in which man perceives himself in relation to that world.

I do not claim to have answers to the above questions. However, I do not see any good reason why, given that man was produced by nature, that there should be a great chasm between the attributes of man and world. This doesn't mean that man cannot have unique attributes, just as a human eye has attributes that a human finger lacks. It is possible that the high-level of thought that occurs in a human is a unique attribute of man. But, just as the body of man is pervaded by unique attributes, and just as the attributes of mercury are very different to those of helium, uniqueness doesn't entail a fundamental division in reality. It is surely the case that the vast majority of the attributes of man are shared by the entire natural world, whilst every phenomenon in the natural world (including man) also has some kind of uniqueness when analysed in detail.

It could be argued that the question: "How much of man is natural?" should be replaced with the question: "How many of the attributes of man are not present in the non-human world?" At a first glance this question would seem to provide some kind of an answer to the former question. If it was concluded that man has a plethora of attributes that are not present in the non-human world then this would seem to imply that a large proportion of man is not natural. Whilst, contrarily, if it was concluded that there are hardly any 'unique' human attributes this would indicate that man is very largely natural. However, it is clear that the latter question cannot provide any kind of answer to the former question. Whilst the latter question is a valid question to ask it is comparable to asking the question: "How many of the attributes of mercury are not present in the non-mercury world?" There clearly are attributes of mercury that are not present in the non-mercury world, because it is the presence of these attributes that makes mercury mercury. But it would be nonsensical to conclude from this that the question: "How much of mercury is natural?" is a sensible question to ask. It is simply the case that within the natural world there are differences in the attributes of the various phenomena that exist; some phenomena will closely resemble others, and some will not.

It is time to consider the place of non-human animals. We have seen that 'natural' is defined as 'present in or produced by the material world and its phenomena', and is conceptually opposed to both the 'artificial' and the 'supernatural'. Given these 'oppositions' what are we to make of the 'naturalness' of other animals. We know that all living things modify their surrounding environment, and that many species of animals are very human-like in their activities. For example, chimpanzees are tool users, beavers construct dams, and birds construct nests. These activities and modifications of the surrounding world by non-human animals are clearly not 'artificial', because artificial is defined as 'made by humans; produced rather than natural'. They are surely also not 'supernatural', which is defined as: 'of or relating to existence outside the natural world'. These activities are surely wholly natural.

I take it that it would be indefensible to describe non-human animals themselves, or their constructions, as anything other than wholly 'natural'. But I also take it that it would be woefully inadequate to describe beavers and chimpanzees as 'mere matter'; there are close links between the attributes of humans and the attributes of these non-human animals. But, it has been accepted that beavers and chimpanzees, and their activities, are wholly 'natural'. This means that they are wholly produced by or present in the material world; they are either 'mere matter' or the result of the interactions of 'mere matter'. So, as with humans, there is a clear tension here – beavers and chimpanzees are both 'natural' *and* they are 'more than mere matter' at the same time.

This tension gets to the heart of the issue of the relationship between man and the 'natural world'. In fact, other animals are an 'intermediary' between man and phenomena such as mercury. It seems easy to assert that mercury is both wholly natural and 'mere matter'. But when it comes to an animal such as a chimpanzee, whilst it is easy to assert that it is wholly natural, it also seems to be correct to assert that it is 'more than mere matter'. This tension gets continued into the realm of man, because man is also surely 'more than mere matter'. It is this tension which leads to the conclusion that maybe man is not wholly 'natural' because the 'natural' is fundamentally 'mere matter'. But if we accept this conclusion then we surely also have to accept that some non-human animals are also not wholly 'natural'. This is surely unacceptable.

What is the alternative? If the similarities between man and certain species of non-human animals are accepted, which they surely should be, then it has to be accepted that if these non-human animals and their activities are wholly natural, then so is man and his activities. Furthermore, we should change our conceptualization and definition of 'natural' by ridding it of the notion of the 'material' world. In other words, we should initially accept our ignorance about the fundamental nature of reality, and then we should conclude that the word 'material' is vacuous. This means that we can then define 'natural' as: *'present in or produced by the world and its phenomena'*. This definition quite helpfully rids us of the notion of the 'supernatural'. It also leaves open the possibility that there is a tight coupling between the attributes of man and world. If this conceptualization became the dominant view of the man-nature relationship, rather than the 'western view', then surely man would consider himself to be wholly natural. In the present epoch man doubts his 'naturalness'.

Why does man doubt his 'naturalness'?

It is slightly paradoxical that man can on the one hand talk of the evolution of all species from a common ancestor and the Big Bang, and yet, on the other hand, he can doubt his 'naturalness'. Perhaps this is so because man is that part of nature which *of necessity* considers itself to be not natural. In other words, in the form of man, nature has produced a kind of 'reflective mirror' which enables nature to do things that are impossible without such a mirror. A useful analogy is the hairdresser, who is capable of creating the perfect haircut for her clients without needing to use a mirror, but who can only produce a dreadful mess on her own hair without the aid of a mirror. If the hairdresser acquires a mirror she will gain the ability to perfectly cut her own hair, just as nature clearly gains abilities through producing man.

If this is right – if man has unique abilities in nature due to not considering himself to be natural – then this means that it is inevitable that man will doubt his naturalness. To be man *is* to consider oneself to be not natural; to be opposed to the surrounding world; to be alienated from the rest of nature. On this view it is inevitable that man will doubt his 'naturalness'.

Concluding remarks

I have argued that man is completely and utterly natural in every fibre of his being. The word 'artificial' is simply a label that is used in human communication to refer to the productive activities of man in the world. The word 'supernatural' simply delineates man's understanding of the world from the way the world actually is.

I have claimed that some species of non-human animals are sufficiently similar to man that if man isn't wholly natural then this implies that these non-human species are also not wholly natural. I have also claimed that it would be futile to attempt to divide the phenomena of the Earth up into the 'natural' and the 'artificial'; the inextricability of the 'artificial'/'natural' division forces one to accept that everything is natural.

I have proposed that a more perplexing question than: "How much of man is natural?" is, "How could man possibly doubt that he is wholly natural?" I have suggested that man is that part of nature which inevitably comes to consider itself to be not natural. It is this belief, this doubt, which gives man his unique position in nature. Man and doubt are inextricably linked. Nevertheless, man is wholly and utterly natural.

Friedrich Hölderlin and the Environmental Crisis

This article contains the first philosophy paper that I ever wrote. It was part of my MA in philosophy, and it was also published in the journal *Cosmos and History: The Journal of Natural and Social Philosophy* in 2007 (Vol. 3, No. 1). There are obvious similarities between my philosophy and Hölderlin's philosophy and I believe that this paper will help you to gain a fuller understanding of my philosophy. However, you should keep in mind that this paper isn't my philosophy; it is an attempt to convince the reader that *Hölderlin would agree with my philosophy*. As Hölderlin is a well-known philosopher, I thought that establishing that he would agree with the central tenet of my philosophy was a worthwhile endeavour. I have a great respect for Hölderlin's writings and feel a close affinity with them; however, we don't agree on absolutely everything. So, if there are contradictions between the views expressed in this paper, and the views expressed in the rest of this book, then this is because Hölderlin and I do occasionally go our separate ways!

***

The German Romantic Friedrich Hölderlin developed a unique perspective on the relationship between humankind and the rest of nature. He believed that humanity has a positive role to play in cosmic evolution, and that modernity is the crucial stage in fulfilling this role. In this article I will be arguing for a reinterpretation of his ideas regarding the position of humankind in cosmic evolution, and for an application of these ideas to the 'environmental crisis' of modernity. This reinterpretation is of interest because it entails an inversion of the conventional notion of causality in the 'environmental crisis'; instead of humans 'harming' nature, in the reinterpretation it is nature that causes human suffering.

Hölderlin's ideas are of particular interest because he yearned for an end to human suffering, but was also firmly convinced that humankind was inevitably destined to be separated from nature, and thereby destined to endure suffering. Hölderlin's conception of the human-nature relationship as part of an unfolding process of cosmological change seems to be of great relevance today, an age that is characterized by belief in the meaninglessness of human existence, and by concern about the way that we have altered the pre-human conditions of the Earth. Hölderlin's views provide a unique perspective on modernity that is worthy of serious consideration.

I start by outlining Hölderlin's views on the role of humankind in universal evolution. I then review the secondary literature on Hölderlin that relates to these ideas. I proceed to argue that Hölderlin's philosophy is applicable to, and gives a unique perspective on, the 'environmental crisis' of modernity. I argue that the existing secondary literature on Hölderlin has not recognized this, and that a reinterpretation of the role of humanity in Hölderlin's philosophy of cosmic evolution is therefore required. My central claim is that for Hölderlin, modernity and the related notion of the contemporary 'environmental crisis' is a necessary stage of cosmic evolution, and thus that it is far from a 'crisis'. Rather it is a necessary stage of disharmony that will inevitably be followed by a re-conquered harmony. I will argue that for Hölderlin this disharmony is characterized by the environmental changes that are resultant from the development of technology.

1. Hölderlin's philosophy of human nature, cosmic evolution and modernity

The starting point of Hölderlin's philosophy is that there must be a basic unknowable reality which precedes self-consciousness wherein subjects and objects are not in existence but are both part of a 'blessed unity of being'. He describes this unity as, "Where subject and object simply are, and not just partially, united...only there and nowhere else can there be talk of being."[1] He argues that the 'blessed unity of being' (which he also refers to as 'nature') is responsible for the coming into existence of humanity through using its power to initiate a division of itself into subjects and objects. This division of being causes the emergence of judgement. Hölderlin states that, "'I am I' is the most fitting example of this concept of judgement...[as] it sets itself in opposition to the *not-I*, not in opposition to *itself*."[2]

The division means that human beings are not capable of actions that are independent of nature; Hölderlin states that, "all the streams of

---

[1] Friedrich Hölderlin, 'Being Judgement Possibility', in J. M. Bernstein (ed.), *Classic and Romantic German Aesthetics,* Cambridge, Cambridge University Press, 2003, p. 191.
[2] Ibid., p. 192.

human activity have their source in nature."³ It is revealing to compare this claim with the words of Hölderlin's character Hyperion, "What is man? – so I might begin; how does it happen that the world contains such a thing, which ferments like a chaos or moulders like a rotten tree, and never grows to ripeness? How can Nature tolerate this sour grape among her sweet clusters?"⁴ For Hölderlin, man is the 'violent' being, whose coming into existence in opposition to the rest of nature was *initiated* by nature.

Hölderlin sees this opposition between man and the rest of nature as culminating in modernity – an era that he claims is characterised by the absence of the gods. In *Brot und Wein* Hölderlin writes, "Though the gods are living, Over our heads they live, up in a different world...Little they seem to care whether we live or do not."⁵ A key question for Hölderlin is how we deal with this separation. He envisions two possibilities – the 'Greek' response which is to dissolve the self and die, and the 'Hesperian' response of a living death.

Hölderlin came to view the 'Greek' response as hubristic, it being based on an anthropocentric (human-centered) desire to oppose the division initiated by nature. He thus sees the 'Hesperian' response of living and carrying out actions that are dependent on nature for their origination as the appropriate non-hubristic response to our separation. Hölderlin's position is that as nature created the

---

³Alison Stone, 'Irigaray and Hölderlin on the Relation Between Nature and Culture', in *Continental Philosophy Review,* vol. 36, no. 4, 2003, p. 423.
⁴Friedrich Hölderlin ,'Hyperion', in Eric L. Santner (ed.), *Hyperion and Selected Poems,* New York, Continuum, 1990, p. 35.
⁵Ibid., p. 185.

separation, *only* nature can bring the separation to an end. He sees this process of separation and reconnection as part of a broader cosmic picture wherein nature is an unfolding organism rather than a huge mechanism. This organismic view enables him to envision teleological processes in nature which give rise to his claim that there will be, "eternal progress of nature towards perfection."[6]

2. Interpretations of Hölderlin and his concept of fate

In this section I set out my view of Hölderlin's conception of fate – that all human actions are part of the evolution of nature towards perfection. I do this by reviewing the existing scholarly literature on Hölderlin and showing that whilst these interpretations all recognise parts of Hölderlin's conception of fate that they do not capture the whole of it. I start with interpretations of human nature, move on to cosmic processes, and finally consider the role of modernity within these processes.

At the level of the human there is a general consensus in the literature that Hölderlin's position is that humans are endowed by nature with qualities that shape human nature, and that this inevitably shapes human interactions with the rest of nature. There are various names in the literature for the qualities which are endowed to humans. Dennis J. Schmidt refers to the qualities present in humans as their 'formative drive.' He claims that, "Hölderlin suggests that human nature and practices are to be understood by reference to a formative drive which expresses itself as a

---
[6]Ronald Peacock, *Hölderlin,* London, Methuen & Co. Ltd, 1938, p. 36.

constant need for 'art'."[7] In a similar vein, Thomas Pfau argues for an 'intellectual intuition.' He states that, "Hölderlin recasts the convergence of "freedom and necessity" as the most primordial synthesis of intellect and intuition itself, a synthesis which takes place within the subject itself. He thus approaches what Kant had repeatedly ruled out as an "intellectual intuition"."[8]

In agreement with Schmidt and Pfau, Franz Gabriel Nauen argues that for Hölderlin, "all men do in fact have the same basic character...all human activity can be derived from the same *elemental drive* in human nature."[9] The 'formative drive' / 'intellectual intuition' / 'elemental drive' identified in the literature reveals why man can be seen as the 'violent' being. Human nature is to engage in 'art', to utilize the resources of nature so that culture can be generated and sustained. This generation of human culture actually benefits nature as a whole, but it requires large-scale modification of parts of non-human nature. The destiny of man is thus a disruptive one. It is clear that it is also an undesirable one. Nauen states that for Hölderlin, "Even war and economic enterprise serve to fulfil the destiny of man, which is to "multiply, propel, distinguish and mix together the life of Nature"."[10]

---

[7] Dennis J. Schmidt, *On Germans and Other Greeks*, Indiana University Press, 2001, p. 139.

[8] Thomas Pfau, *Friedrich Hölderlin: Essays and Letters on Theory*, New York, SUNY Press, 1988, p. 15.

[9] Franz Gabriel Nauen, *Revolution, Idealism and Human Freedom: Schelling, Hölderlin and Hegel and the Crisis of Early German Idealism*, Indiana University Press, 2001, p. 139.

[10] Ibid.

So Hölderlin sees human nature, economic production and even war as parts of a broader cosmic evolutionary process; the universe *as a whole* is seen as evolving to perfection. There will inevitably be aspects of this evolution that from a narrow perspective could be viewed as 'less than perfect'. These negative aspects of the evolutionary process – from war, to the presence of evil in its entirety – have to be seen as inescapable parts of the whole process.

The key point is that for Hölderlin the cosmic evolutionary process *ends* in perfection. Thus, Ronald Peacock argues that, "the division produced by conflict is followed by a re-conquered harmony."[11] Similarly, Anselm Haverkamp argues that an interpretation of the poems *Andenken* and *Mnemosyne* is the expression, 'where danger threatens, salvation also grows.'[12] Whilst, Martin Heidegger translates the opening lines of *Patmos* as, "But where danger is, grows the saving power also."[13] Hölderlin's view is clearly that from a narrow and short-term perspective danger and conflict are often the norm, but that these things actually play a part in bringing about a greater harmony in the future. In the long-term they are all part of the evolution of the whole universe to perfection.

Cosmic evolution is thus one long process of disharmonies and inevitably following harmonies. Peacock argues that Hölderlin's vision

---

[11] Peacock, *Hölderlin*, p. 22.

[12] Anselm Haverkamp, *Leaves of Mourning: Hölderlin's Late Work,* New York, SUNY Press, 1996, p. 48.

[13] Martin Heidegger, 'The Question Concerning Technology', in R.C. Scharff and V. Dusek (eds.), *Philosophy of Technology: The Technological Condition – An Anthology,* Oxford, Blackwell Publishing, 2003, p. 261.

is of a, "harmonised process of life which comprises within itself the rhythmic movement from chaos to form and back again, and an emotional experience of this which in the sphere of nature knows only the one rapture, but in the human sphere suffering and joy."[14] It is revealing that this interpretation sees 'violent' humans as suffering, whilst nature is purely rapturous. This clearly sheds light on the question posed by Hölderlin's character Hyperion: "How can Nature tolerate this sour grape among her sweet clusters?"[15] The answer seems to be that human 'violence' *enables* nature to be rapturous. As part of this rapture humans experience suffering.

Why should suffering be a uniquely human experience? To explain this Peacock cites part of a letter from Hölderlin to his brother, "Why can they [humans] not live contented like the beasts of the field? he asks: and replies that this would be as unnatural in man, as in animals the tricks, or arts, man trains them to perform. Thus he establishes that the arts of man are natural to man. Culture, then, derives from nature; and the impulse to it is the characteristic which distinguishes man from the rest of creation."[16]

The human impulse to culture has culminated in the era of modernity. Hölderlin sees this period as one of great significance as he sees it as a historical epoch that is characterised by the *absence of the gods.* To be consistent with his views on harmonised evolution to perfection there must be a reason for this absence. Indeed, Peacock

---

[14]Peacock, *Hölderlin*, p. 22.
[15]Hölderlin, 'Hyperion', p. 35.
[16]Peacock, *Hölderlin*, p. 36.

argues that Hölderlin thinks that, "a godless age is part of a divine mystery, it is as necessary as day, ordained by a higher power."[17] Furthermore, Heidegger claims that the gods are still present, despite their absence: "man who, even with his most exulted thought could hardly penetrate to their Being, even though, with the same grandeur as at all time, they were somehow there."[18]

The absence of the gods in modernity is deeply related to the contemporary danger that exists in modernity. It should be remembered that this danger cannot be a cause for concern for Hölderlin – as all dangers are inevitably followed by regained harmonies. Nevertheless, Heidegger attempts to identify the exact danger that Hölderlin believed is present in modernity. Heidegger claims that, "the essence of technology, enframing, is the extreme danger."[19] It must follow that for Heidegger, "precisely the essence of technology must harbor in itself the growth of the saving power."[20] He sees this as occurring when the essential unfolding of technology gives rise to the possibility of opening up a "free relation" with technology which is inclusive of non-instrumental possibilities.[21]

In an interpretation of the 1802 hymn *Friedensfeier*, Richard Unger draws out Hölderlin's views on the absence of the gods in

---

[17] Ibid., p. 92.
[18] Martin Heidegger, *Existence and Being*, London, Vision Press Ltd., 1956, p. 190.
[19] Heidegger, 'The Question Concerning Technology', p. 261.
[20] Ibid.
[21] R.C. Scharff and V. Dusek, 'Introduction to Heidegger on Technology', in R.C. Scharff and V. Dusek (eds.), *Philosophy of Technology: The Technological Condition – An Anthology*, Oxford, Blackwell Publishing, 2003, p. 248.

modernity.[22] In *Friedensfeier* the entire span of Western civilization is characterised as a thunderstorm which is ruled by a "law of destiny" which ensures that a certain amount of "work" is achieved. Unger argues that it is clear that this "work", "is the product of the storm itself and that it designates the harmonious totality of earthly existence during the coming era."[23] The end of the "storm" of modernity enables the arrival of a mysterious "prince" who makes it possible that, "men can now for the first time hear the "work" that has been long in preparation "from morning until evening"."[24]

Following the inevitable successful accomplishment of the "work" of Western civilization, the great Spirit will disclose a Time-Image which will, "be a comprehensive depiction of the historical process and its triumphant result."[25] Unger argues that, "the Image shows that there is an alliance between the Spirit of history and the elemental divine presences of nature – for the natural elements with which man has always worked have played integral and essential parts in man's history."[26] The triumphant result of the actions of humankind in modernity is clearly an example of a re-conquered harmony that follows division.

In Unger's interpretation of *Friedensfeier* we have a picture of modernity in which humans are carrying out "work" under a "law of destiny". The crucial factor is that humanity is ignorant that it is

---

[22] Richard Unger, *Friedrich Hölderlin,* Boston, Twayne Publishers, 1984, pp. 100-105.
[23] Ibid., p. 102.
[24] Ibid., p. 101.
[25] Ibid., p. 104.
[26] Ibid., p. 105.

working under a "law of destiny" in modernity, until modernity has ended. It is then that through the Time-Image the great Spirit reveals the successful outcome of modernity, and the *nature and value* of the accomplished "work". This is a prime example of a short-term and narrow perspective entailing the perception of a lack of destiny and of needless suffering, whilst in the longer-term the same events are seen to be an inevitable part of a broader positive outcome – the evolution of the universe to perfection.

This difference of perspectives can explain an apparent contradiction in the literature between Unger's interpretation of *Friedensfeier,* and Schmidt's analysis of Hölderlin's 1801 letter to Bohlendorff. This letter was written only one year before *Friedensfeier* and Schmidt claims that in it Hölderlin's position is, "that the peculiar flow of modernity is the lack of destiny."[27] The apparently contradictory views of Unger and Schmidt can be reconciled through recalling Peacock's interpretation that, "a godless age is part of a divine mystery, it is as necessary as day, ordained by a higher power,"[28] and comparing it to Unger's claim that men are blind to the point of the "work" that they have been carrying out until the "storm" of Western civilization has passed.

The comparison reveals that the "law of destiny" applies to the activities of *humanity as a collective* in Western history, activities that are ordained by a higher power for a specific purpose. In contrast, the "lack of destiny" applies to *individual human beings.* This difference

---

[27] Schmidt, *On Germans and Other Greeks,* p. 137.
[28] Peacock, *Hölderlin,* p. 92.

arises because individual humans are unaware that their actions are part of an inevitably unfolding cosmic plan, it is only the fruition of the plan that enables realization. Instead, humans believe that they have free will and live in a meaningless age. Therefore, modernity can at one and the same time be characterized as both a period governed by a "law of destiny" and a period constituted by a "lack of destiny". The difference is purely one of perspective.

This conception of modernity as simultaneously being a period of a "lack of destiny" and a "law of destiny" raises the issue of human attitudes to nature. If human attitudes and actions towards nature are in the interests of nature, then it seems that there is no such thing as a truly 'human-centered' attitude. The appropriate attitude that humans should take to the objective side of nature, given Hölderlin's philosophy, has been addressed by Alison Stone. She argues that because, "according to Hölderlin's thinking, we have become separated from nature by *its* power alone, so it is not within *our* power to undo separation."[29] Therefore, "the appropriately modest response is to endure separation – to wait, patiently, until nature may change its mode of being."[30] This means that the appropriate human attitude entails, "the *acceptance* of disenchantment, of separation, of meaninglessness."[31]

---

[29] Stone, 'Irigaray and Hölderlin on the Relation Between Nature and Culture', p. 424.
[30] Ibid.
[31] Alison Stone, *Nature in Continental Philosophy – Week 4, Section V, Friedrich Hölderlin*, [online], http://www.lancaster.ac.uk/depts/philosophy/awaymave/408new/wk4.htm, [accessed 25 October 2005].

This view is concordant with the "lack of destiny" perspective. However, when the "law of destiny" is taken into account, then the hidden meaning is revealed. Furthermore, the whole notion of the attitudes of individual humans then becomes irrelevant. It seems that there cannot be such a thing as a *truly* human-centered attitude, because all attitudes originate from nature, and they all lead to actions which fulfil the "law of destiny". It may seem that our attitudes to nature are of importance, but this is because we believe in a "lack of destiny", and are inevitably blind to the bigger picture of the "law of destiny". Whatever our attitudes as individuals, our relationship with the rest of nature as a collective would be 'for the best'.

3. A reinterpretation of the human in cosmic evolution

The interpretations of Hölderlin that I have reviewed all give an accurate representation of Hölderlin's views. However, they are all partial views. They all miss the 'big picture' of what Hölderlin's views imply about what it means to be a human in the context of cosmic evolution, and the consequent implications for the perspective from which we should view modernity and the 'environmental crisis'. In an attempt to fully grasp these implications I am going to defend the thesis that: *Hölderlin's philosophy leads to the conclusion that the 'environmental crisis' is a necessary stage in the purposeful evolution of nature towards perfection.* This is an interesting thesis because, if accepted, it would supplant the conception of the meaninglessness of human existence with a conception of positive cosmic purpose.

The argument I will be making centers on three key aspects of Hölderlin's philosophy. Firstly, that he believes that nature is purposefully evolving towards perfection. Secondly, that he believes that the achievement of this perfection requires human actions. Thirdly, that he believes that human actions are determined by nature. Acceptance of these three claims leads to the conclusion that human actions are determined by nature as a necessary stage in the purposeful evolution of nature towards perfection. As the 'environmental crisis' of modernity is purely resultant from human actions, a second conclusion inevitably follows. This is that the 'environmental crisis' itself is determined by nature as a necessary stage in the purposeful evolution of nature towards perfection.

I will now present evidence to support the three key claims. The first claim is that Hölderlin's belief is that *nature is purposefully evolving towards perfection.* The universe can either be viewed as a giant mechanism or as an unfolding organism; Hölderlin clearly held the latter view. This conception of the universe explains his belief that nature unfolds in a way that serves its own purposes; that disharmonies are followed by regained harmonies. This is why Peacock claims that Hölderlin believed in, "the eternal progress of nature towards perfection,"[32] and, "the emergence of perfection in the course of natural development."[33]

---

[32] Peacock, *Hölderlin,* p. 36.
[33] Ibid., p. 105.

This firm belief clashed with Hölderlin's personal yearning for immediate perfection in life. His immense desire to see a morally just world was completely at odds with his philosophical belief that the perfection he sought could only be achieved in the course of natural development. The movement to perfection envisioned by Hölderlin is thus a fatalistic one, an inevitable evolutionary progression towards perfection. Peacock captures this with his claim that for Hölderlin there is an, "acute sense of 'Fate', of inevitability, expressed again and again in his work. Fate is revealed in the process of history… it is inherent in the passage of form to chaos, and of disintegration to a new harmony."[34]

This first claim is the most straightforward of the three. The second claim is that *Hölderlin believes that the achievement of perfection requires human actions.* The starting point in defending this claim is Hölderlin's central belief that nature *used its power* to divide itself and thereby create humankind. This division means that the split was part of the evolutionary process rather than a random occurrence. We can ask ourselves why this may have been a necessary occurrence. An initial answer seems to be Nauen's claim that, "Even war and economic enterprise serve to fulfil the destiny of man, which is to "multiply, propel, distinguish and mix together the life of Nature"."[35]

---

[34] Ibid., p. 93.

[35] Nauen, *Revolution, Idealism and Human Freedom: Schelling, Hölderlin and Hegel and the Crisis of Early German Idealism*, p. 139.

In *The Perspective from which we Have to look at Antiquity* Hölderlin asserts that, "antiquity appears altogether opposed to our primordeal drive which is bent on forming the unformed, to perfect the primordial-natural so that man, who is born for art, will naturally take to what is raw, uneducated, childlike rather than to a formed material where there has already been pre-formed [what] he wishes to form."[36] In a letter to his brother he also asserts that, "the impulse to art and culture…is really a service that men render nature."[37]

The source of Hölderlin's primordeal drive to art is nature, because it is nature that created us and endowed us with our capabilities. This is clear from Peacock's interpretation that, "Man cannot be master of nature; his arts, *necessary though they may be in the scheme of things,* cannot produce the substance which they mould and transform; they can only develop the creative force, which in itself is eternal and not their work."[38]

Hölderlin's primordeal drive to art in humans has inevitably led to the epoch of modernity. Human actions in this epoch appear to be central to the achievement of perfection. Hölderlin claims that modernity is an epoch that, "is as necessary as day, ordained by a higher power."[39] Furthermore, humans have been involved in "work" in modernity that is clearly constitutive of the importance of the

---

[36] Friedrich Hölderlin, 'The Perspective from which We Have to Look at Antiquity', in Thomas Pfau (ed.), *Friedrich Hölderlin: Essays and Letters on Theory,* New York, SUNY Press, 1988, p. 39.
[37] Peacock, *Hölderlin,* p. 37.
[38] Ibid.
[39] Ibid., p. 92.

epoch. This is clear from Unger's interpretation of *Friedensfeier* in which the "law of destiny" ensures that a certain amount of human "work" is done. The crucial factor is that humanity is ignorant that it is working under a "law of destiny" in modernity, until modernity has ended. It is then that through the Time-Image the great Spirit reveals the successful outcome of modernity, and the nature and value of the accomplished "work".

There is no doubt that in Hölderlin's view human actions and their resultant "work" in modernity are part of purposeful evolution to perfection. What is interesting is the exact nature of the "work". There is an obvious connection between the "work" of modernity (*Friedensfeier*) and the "danger" we face in modernity (*Patmos*). Heidegger's interpretation of *Patmos* that, "the essence of technology, enframing, is the extreme danger,"[40] makes it clear that the "work" of modernity is the development of technology. In fact, technological development in modernity seems to be the culmination of Hölderlin's primordeal drive to art. Furthermore, it is very hard to think of any other distinctive aspects of modernity that are resultant from human actions, present an extreme danger, and have cosmic significance. Therefore, for Hölderlin, the achievement of perfection seems to require the human development of technology.

It is interesting that Heidegger sees the danger we face from the "work" of modernity as the essence of technology rather than actual technology. Andrew Feenberg has criticised Heidegger for this abstract concentration on essences rather than the actual

---

[40] Heidegger, 'The Question Concerning Technology', p. 261.

technology itself.[41] A 'Feenberg interpretation' of *Patmos* seems to me to be more in accordance with Hölderlin's views than the 'Heidegger interpretation', as Hölderlin's philosophy is grounded in actualities rather than essences. Hölderlin sees a positive role for actual technology in cosmic evolution; this means that *actual technology* has a cosmic purpose. Therefore, it seems that both the danger we face, and the saviour, must be the *actual* technology developed by human actions.

The importance of the human split from the rest of nature can also be seen in the words of Hölderlin's character *Hyperion*: "How should I escape from the union that binds all things together? We part only to be more intimately one, more divinely at peace with all, with each other. We die that we may live."[42] Human actions are thus depicted as a 'living death' that is necessary for the life (and continued movement to perfection) of nature as a whole. This explains Peacock's interpretation that, "the sphere of nature knows only the one rapture, but in the human sphere [there is] suffering and joy."[43]

The third claim is that *Hölderlin believes that human actions are determined by nature.* There are many passages in Hölderlin's novel *Hyperion* that attribute the responsibilities for human actions to a power or god: "There is a god in us who guides destiny as if it were a

---

[41] Andrew Feenberg, 'Critical Evaluation of Heidegger and Borgmann', in R.C. Scharff and V. Dusek (eds.), *Philosophy of Technology: The Technological Condition – An Anthology*, Oxford, Blackwell Publishing, 2003, pp. 327-337.

[42] Hölderlin, 'Hyperion', p. 123.

[43] Peacock, *Hölderlin*, p. 22.

river of water, and all things are his element."[44]..... "oh forgive me, when I am compelled! I do not choose; I do not reflect. There is a power in me, and I know not if it is myself that drives me to this step."[45]..... "I once saw a child put out its hand to catch the moonlight; but the light went calmly on its way. So do we stand trying to hold back everchanging Fate. Oh, that it were possible but to watch it as peacefully and meditatively as we do the circling stars."[46]..... "Man can change nothing and the light of life comes and departs as it will."[47]..... "We speak of our hearts, of our plans, as if they were ours; yet there is a power outside of us that tosses us here and there as it pleases until it lays us in the grave, and of which we know not where it comes nor where it is bound."[48]

Hölderlin's belief in the lack of human free will is perhaps clearest in his claim in a letter to his mother regarding the views of Spinoza that, "one *must* arrive at his ideas if one wants to explain everything."[49] Spinoza's ideas can be summed up as, "Nature in all its aspects is governed by necessary laws, and human being no less than the rest of nature is determined in all its actions and passions, contrary to those who conceive of it as 'a dominion within a dominion'."[50]

---

[44]Hölderlin, 'Hyperion', p. 11.
[45]Ibid., p. 79.
[46]Ibid., p. 22.
[47]Ibid., p. 127.
[48]Ibid., p. 29.
[49]Friedrich Hölderlin, 'No.41: To his Mother', in Thomas Pfau (ed.), *Friedrich Hölderlin: Essays and Letters on Theory,* New York, SUNY Press, 1988, p. 120.
[50]Moira Gatens, *Imaginary Bodies: Ethics, Power and Corporeality,* London, Routledge, 1996, p. 111.

In order to make abundantly clear Spinoza's – and thus Hölderlin's – views on a lack of human free will here are two quotes from Spinoza: "I say that thing is free which exists and acts solely from the necessity of its own nature...I do not place Freedom in free decision, but in free necessity."[51] And, "a stone receives from an external cause, which impels it, a certain quantity of motion, with which it will afterwards necessarily continue to move...Next, conceive, if you please, that the stone while it continues in motion thinks, and knows that it is striving as much as possible to continue in motion. Surely this stone, inasmuch as it is conscious only of its own effort, and is far from indifferent, will believe that it is completely free, and that it continues in motion for no other reason than because it wants to. And such is the human freedom which all men boast that they possess, and which consists solely in this, that men are conscious of their desire, and ignorant of the causes by which they are determined."[52]

Furthermore, in an interpretation of Hölderlin's *Stutgard,* Peacock argues that, "the laws of growth govern the culture as well as the lives of men...the one process comprehends all things and the one rhythm manifests itself again and again...in the progress of history; in the spiritual life of individuals."[53] In this vision not only human nature, but also the evolution of culture, is seen as an inevitable historical progression. Peacock's interpretation of Hölderlin is that, "man's spirit is but part of the One Spirit,"[54] which Hölderlin insists

---

[51] Benedict de Spinoza, 'LVIII: To Schuller', trans. A. Wolf (ed.), *The Correspondence of Spinoza*, 2nd ed., London, Frank Cass & Co. Ltd., 1966, pp. 294-5.
[52] Ibid., p. 295.
[53] Peacock, *Hölderlin*, p. 25.
[54] Ibid., p. 90.

is involved in a "movement...through successive historical generations."[55] The spirit of man is thus governed by the larger Spirit of nature. This is the sense in which, "all the streams of human activity have their source in nature."[56]

The nature of the relationship between man's spirit and the Spirit of nature is made clear in the following quote from Hölderlin's character Diotima: "a *unique destiny* bore you away to solitude of spirit as waters are borne to mountain peaks."[57] This concept of individual humans having a unique destiny was the view of Johann Herder, who was one of Hölderlin's inspirations. Herder saw nature as a great current of sympathy running through all things which manifested itself in unique inner impulses within different individuals. This means that every human has a unique calling – an original path which they ought to tread. As Herder states, "Each human being has his own measure, as it were an accord peculiar to him of all his feelings to each other."[58] Clearly, for both Herder and Hölderlin, human actions at any one time are determined in accordance with the movements of the One Spirit of nature.

I have presented evidence for the claims that for Hölderlin: *nature is purposefully evolving towards perfection, the achievement of this perfection requires human actions, and human actions are determined by nature.* Acceptance of these three claims leads to the

---

[55] Ibid., p. 114.
[56] Stone, 'Irigaray and Hölderlin on the Relation Between Nature and Culture', p. 423.
[57] Hölderlin, 'Hyperion', p. 122.
[58] Charles Taylor, *Sources of the Self: The Making of the Modern Identity*, Massachusetts, Harvard University Press, 1994, p. 375.

conclusion that human actions are determined by nature as a necessary stage in the purposeful evolution of nature towards perfection. I now briefly argue that the 'environmental crisis' of modernity is purely resultant from human actions.

The definition of an environmental problem is: "any change of state in the physical environment which is brought about by human interference with the physical environment, and has effects which society deems unacceptable in the light of its shared norms."[59] This definition encapsulates a sliding scale of environmental problems from those that are local and temporary on the one hand, to those that are global and long-lasting on the other. The 'environmental crisis' as a concept has arisen because of the emergence in the last 100 years of an increasing number of environmental problems that are towards the global and long-lasting end of the scale. The 'environmental crisis' is thus purely resultant from the *human actions* which have created environmental problems that are characterised by their global reach and long-lasting nature.

This means that the above conclusion, that human actions are determined by nature as a necessary stage in the purposeful evolution of nature towards perfection, needs amending. As the 'environmental crisis' is purely resultant from human actions, it too must be part of this purposeful evolution. Therefore, the new conclusion that inevitably follows is:

---

[59]Peter B. Sloep and Maris C.E. van Dam-Mieras, 'Science on Environmental Problems', in P. Glasbergen and A. Blowers (eds.) *Environmental Policy in an International Context: Perspectives,* Oxford, Butterworth-Heinmann, 2003, p. 42.

- *The 'environmental crisis' is determined by nature as a necessary stage in the purposeful evolution of nature towards perfection.*

4. Objections to the reinterpretation

It could be objected that there are many references to human freedom in Hölderlin's work that would seem to cast doubt on the third claim. This is particularly noticeable in his novel Hyperion. For example, Hyperion states that, "without freedom all is dead."[60] However, this objection is easily answered because these references all appear in Hölderlin's early work, and even then they are more than counterbalanced by the opposing fatalistic views that I have outlined. In his early period Hölderlin was struggling to come to terms with the conflict between his keen moral aspirations for social change on the one hand, and his belief in perfection only arising through natural development on the other. In his later work, as is clear in his endorsement of the 'Hesperian' response to our condition, he firmly accepts the powers of natural development and the determination of human actions by nature. He realizes the futility of pursuing his idealistic moral aspirations because he accepts the illusory nature of human free will.

A further objection could be made that this reinterpretation is pointless because Darwin's theory of evolution, which emerged shortly after Hölderlin's time, gives a view of evolutionary processes

---

[60] Hölderlin, 'Hyperion', p. 117.

that is incompatible with Hölderlin's view that there was a 'blessed unity of being' prior to the arrival of humans. We now know that the emergence of the human species – and its primordeal drive to art – was preceded by four billion years of evolution of life on Earth. It can thus be argued that there was not a 'blessed unity of being' prior to the evolution of humankind.

This is exemplified by the claim of Hans Jonas that the subject-object divide opened up four billion years ago, when, "living substance, by some original act of segregation, has taken itself out of the general integration of things in the physical context, set itself over against the world, and introduced the tension of "to be or not to be" into the neutral assuredness of existence."[61] This certainly does not appear to be a pre-human 'blessed unity of being'. However, it is interesting that Jonas also sees humans as, "a 'coming to itself' of original substance."[62]

It is clear that this Darwinian based objection does not invalidate the views of Hölderlin, or the reinterpretation of them presented here. In fact, not only does evolutionary theory perfectly complement Hölderlin's philosophy, his philosophy *needs* it. The idea that nature could use its power to instantaneously create a being as complex as a human out of the 'blessed unity of being' is hardly defensible. In the light of our knowledge today we can simply reinterpret Hölderlin as claiming that nature used its power four billion years ago to divide

---

[61] Hans Jonas, *The Phenomenon of Life: Toward a Philosophical Biology*, Illinois, Northwestern University Press, 2001, p. 4.
[62] Ibid., p. xv.

the 'blessed unity of being' and create a subject/object divide. As he sees nature as an unfolding and evolving organism, the divide would give rise to human subjects after a sufficient period of time. This, ""coming to itself" of original substance", as Jonas describes it, has in actuality taken approximately four billion years.

5. Conclusion

I have argued that the existing secondary literature has not grasped the full implications of Hölderlin's thought for what it means to be a human in modernity. By drawing together Hölderlin's ideas I have sought to understand his notion of the purpose of human actions, and what this purpose means for the 'environmental crisis'.

Hölderlin's conception of nature is an organism unfolding to perfection. I have argued that he sees modernity as an important stage of this unfolding, which is characterized by the development of technology through human actions. I have further argued that this means that the 'environmental crisis' of modernity – a side-effect of the development of technology – is also an inevitable stage of this unfolding; it is in the interests of nature. As nature continues to unfold, the disharmony of modernity will be succeeded by a re-conquered harmony. I have argued that Hölderlin's 'saving power' is actual technology, as this seems most consistent with his thought. Heidegger's view, that the 'saving power' is the essencing of technology, seems inconsistent with the positive role of technology in cosmic evolution that is envisioned by Hölderlin.

The reinterpretation I have outlined clearly entails an inversion of the conventional notion of causality in the 'environmental crisis' of modernity. Humanity is conventionally pictured as harming nature. My thesis has shown that for Hölderlin it is nature that is 'harming' humanity. We have been cast aside out of the rapture of nature into a realm of suffering and self-consciousness, with the purpose of developing technology to serve the purposes of the unfolding nature of which we are a part.

We are left with the question of what our attitudes to nature should be, given this reinterpretation of what it means to be a human in cosmic evolution. The answer is simple. As nature is the source of our individual attitudes, our attitudes to nature must be in the interests of nature. Our attitudes, whether they are techno-centric, environmentalist, quietist, or nature-exploitative are all correct for us as individuals, because in the aggregate they fulfil the purpose of nature as a whole.

BIBLIOGRAPHY

Feenberg, Andrew, 'Critical Evaluation of Heidegger and Borgmann', in R.C. Scharff and V. Dusek (eds.), *Philosophy of Technology: The Technological Condition – An Anthology*, Oxford, Blackwell Publishing, 2003.

Gatens, Moira, *Imaginary Bodies: Ethics, Power and Corporeality*, London, Routledge, 1996. Haverkamp, Anselm, *Leaves of Mourning: Hölderlin's Late Work*, New York, SUNY Press, 1996.

Heidegger, Martin, *Existence and Being*, London, Vision Press Ltd., 1956.

Heidegger, Martin, 'The Question Concerning Technology', in R.C. Scharff and V. Dusek (eds.), *Philosophy of Technology: The Technological Condition – An Anthology*, Oxford, Blackwell Publishing, 2003.

Hölderlin, Friedrich, 'Being Judgement Possibility', in J. M. Bernstein (ed.), *Classic and Romantic German Aesthetics*, Cambridge, Cambridge University Press, 2003.

Hölderlin, Friedrich, 'Hyperion', in Eric L. Santner (ed.), *Hyperion and Selected Poems*, New York, Continuum, 1990.

Hölderlin, Friedrich, 'No.41: To his Mother', in Thomas Pfau (ed.), *Friedrich Hölderlin: Essays and Letters on Theory*, New York, SUNY Press, 1988.

Hölderlin, Friedrich, 'The Perspective from which We Have to Look at Antiquity', in Thomas Pfau (ed.), *Friedrich Hölderlin: Essays and Letters on Theory,* New York, SUNY Press, 1988.

Jonas, Hans, *The Phenomenon of Life: Toward a Philosophical Biology,* Illinois, Northwestern University Press, 2001.

Nauen, Franz Gabriel, *Revolution, Idealism and Human Freedom: Schelling, Hölderlin and Hegel and the Crisis of Early German Idealism,* Indiana University Press, 2001.

Peacock, Ronald, *Hölderlin,* London, Methuen & Co. Ltd, 1938.

Pfau, Thomas, *Friedrich Hölderlin: Essays and Letters on Theory,* New York, SUNY Press, 1988.

Scharff, R. C., and Dusek, V., 'Introduction to Heidegger on Technology', in R.C. Scharff and V. Dusek (eds.), *Philosophy of Technology: The Technological Condition – An Anthology,* Oxford, Blackwell Publishing, 2003.

Schmidt, Dennis J., *On Germans and Other Greeks,* Indiana University Press, 2001.

Sloep, Peter B., and Dam-Mieras, Maris C.E. van, 'Science on Environmental Problems', in P. Glasbergen and A. Blowers (eds.) *Environmental Policy in an International Context: Perspectives,* Oxford, Butterworth-Heinmann, 2003.

Spinoza, Benedict de, 'LVIII: To Schuller', trans. A. Wolf (ed.), *The Correspondence of Spinoza,* 2nd ed., London, Frank Cass & Co. Ltd., 1966.

Stone, Alison, 'Irigaray and Hölderlin on the Relation Between Nature and Culture', in *Continental Philosophy Review*, vol. 36, no. 4, 2003. Stone, Alison, *Nature in Continental Philosophy – Week 4, Section V, Friedrich Hölderlin*, [online], http://www.lancaster.ac.uk/depts/philosophy/awaymave/408new/wk4.htm
[accessed 25 October 2005].

Taylor, Charles, *Sources of the Self: The Making of the Modern Identity*, Massachusetts, Harvard University Press, 1994.

Unger, Richard, *Friedrich Hölderlin*, Boston, Twayne Publishers, 1984.

Friedrich Hölderlin: A Final Reflection

In the last article I claimed that Hölderlin believes that technology has a positive role to play in cosmic evolution; technology presents a danger but it is also fundamentally a 'saving power'. Given that Hölderlin lived before the human species formulated the concept of the 'environmental crisis', before there was an awareness of non-human-induced global warming, before there was an awareness of human-induced global warming, one cannot reasonably expect him to have worked out exactly why technology is simultaneously a 'danger' and a 'saving power'.

Technology is a 'danger' because its unleashing results in the release of massive amounts of fossil fuels from underground storage; this release of fossil fuels into the biogeochemical cycles of the Earth perturbs the planetary homeostatic regulatory capacity at a time when it is already severely weakening.

Technology is a 'saving power' because it is the 'armour' which life on Earth needs in order to survive and thrive into the distant future. The central component of this 'technological armour' is the technology which will be deployed to regulate the temperature of the Earth's atmosphere. This deployment is required because solar systems age and this means that it is inevitable that the homeostatic regulatory capacity of the planet that one inhabits will weaken in the face of non-human-induced global warming.

The Philosophy of Global Warming

As a final reflection it is worth recalling Hölderlin's belief that:

> The end of the "storm" of modernity enables the arrival of a mysterious "prince" who makes it possible that, "men can now for the first time hear the "work" that has been long in preparation "from morning until evening".
>
> (p. 382 of this book)

Could the "storm of modernity" have ended?

Could this book enable men to "hear for the first time" the purpose of their "work"?

Could I be a "mysterious prince"?!

Further Reading

Books by the Author:

I initially outlined my philosophical worldview four years ago. The book you are now reading is much more comprehensive, but you still might find some things of interest in the original work:

Is the Human Species Special?: Why human-induced global warming could be in the interests of life, Cranmore Publications, 2010.

If you are interested in learning more about the nature of reality, the nature of mind, the philosophy of biology and my 'panwhat-it-is-likeness' view, then you need this book:

An Evolutionary Perspective on the Relationship Between Humans and Their Surroundings: Geoengineering, the Purpose of Life & the Nature of the Universe, Cranmore Publications, 2012.

If you are particularly keen to read more of what I have written then you could look at these fairly short books which illuminate particular aspects of my philosophy:

Saviours or Destroyers: The relationship between the human species and the rest of life on Earth, Cranmore Publications, 2012.

What Does it Mean to be 'Green'?: Sustainability, Respect & Spirituality, Vitae Publications, 2011.

Should I be a Vegetarian?: A personal reflection on meat-eating, vegetarianism and veganism, Cranmore Publications, 2011.

Other Worthwhile Books:

If you are interested in learning more about the Earth's homeostatic regulatory capacity then you can look up the work of Sir James Lovelock:

Gaia: A New Look at Life on Earth, OUP, 1982.

The Ages of Gaia: A Biography of Our Living Earth, OUP, 1995.

The Revenge of Gaia, Penguin Books Ltd., 2006.

If you are interested in learning more about the directionality of human culture from hunter gatherer to globalised technological society, then it is worth reading Robert Wright's book:

Nonzero: History, Evolution and Human Cooperation, Abacus, 2000.

If you are interested in learning more about the correspondences between the alignments of the outer planets of the Solar System and the evolutionary progression of human culture, then the following book by Professor Richard Tarnas is for you:

Cosmos and Psyche: Intimations of a New World View, Plume, 2007.

Keeping in Contact

You can find new articles on my blog:

http://www.neilpaulcummins.blogspot.co.uk

You can find out where to get more copies of this book here:

http://www.cranmorepublications.co.uk/pogw

You can also email me if you have a question or a comment:

neilpaulc@gmail.com

Lightning Source UK Ltd.
Milton Keynes UK
UKOW06f2130290315

248717UK00002B/38/P